The Logica Yearbook
2011

The Logica Yearbook 2011

Edited by
Michal Peliš
and
Vít Punčochář

© Individual author and College Publications 2012.
All rights reserved.

ISBN 978-1-84890-071-4

College Publications
Scientific Director: Dov Gabbay
Managing Director: Jane Spurr
Department of Computer Science
King's College London, Strand, London WC2R 2LS, UK

www.collegepublications.co.uk

Original cover design by Laraine Welch
Printed by Lightning Source, Milton Keynes, UK

All rights reserved. No part of this publication may be reproduced, stored in a retrieval system or transmitted in any form, or by any means, electronic, mechanical, photocopying, recording or otherwise without prior permission, in writing, from the publisher.

Preface

Since 1997 participants in the Logica symposia have had the opportunity of publishing their contributions in The Logica Yearbook series. Last year was no exception and so we have the pleasure of introducing the latest volume of the proceedings which contains most of the papers presented at Logica 2011.

The international symposium Logica has a long and rich tradition and, in this respect, last year was very special because we were celebrating an important anniversary. Logica 2011 was the 25th event in the series of conferences annually held in the Czech Republic.

Logica 2011, held at Hejnice Monastery (North Bohemia) from 20th to 24th June 2011, was organized by the Department of Logic in the Institute of Philosophy of the Academy of Sciences of the Czech Republic. As every year, the symposium brought together logicians from the whole world and besides the invited talks (invited speakers were Edwin Mares, Pavel Materna, Krister Segerberg and Gila Sher) about thirty other papers devoted to the various branches of logic were presented.

Both the Logica symposium and The Logica Yearbook are the result of the joint effort of many people who deserve our warmest thanks. We thank Vladimír Svoboda, the head of the Organizing Committee of Logica 2011. We are very grateful to the Institute of Philosophy and especially its director, Pavel Baran, for all their support. We would like to thank College Publications and its managing director Jane Spurr. We greatly appreciate the hard work of Karel Chvalovský, the typesetter of this volume. Special thanks go to Petra Ivaničová who provided invaluable assistance to the organizers of the conference.

We are also very grateful to the staff of Hejnice Monastery and to Bernard Family Brewery of Humpolec which has traditionally sponsored the social programme of the symposium. Neither the publication of this volume, nor the conference Logica 2011 itself would be possible without the Grant Agency of the Czech Republic which provided significant support by financing the grant project no. 401/04/0117.

Last, but not least, we would like to thank all the authors for their exemplary collaboration during the editorial process.

Prague, May 2012   Michal Peliš and Vít Punčochář

Contents

A Unifying Framework for Reasoning about Normative Conflicts   1
  Mathieu Beirlaen

Non-Normal Worlds and Representation   15
  Francesco Berto

The Evaluation Semantics—A Short Introduction   31
  Frode Bjørdal

On Dialetheic Entailment   37
  Massimiliano Carrara, Enrico Martino, and Vittorio Morato

The Asymmetry of Formal Logic   49
  Colin Cheyne

Reasoning by Analogy in Inductive Logic   63
  Alexandra Hill and Jeff Paris

The Confirmation of Singular Causal Statements by Carnap's
Inductive Logic   77
  Yusuke Kaneko

Logics of Fact and Fiction, Where Do Possible Worlds Belong?   97
  John T Kearns

How to Build a Deontic Action Logic   107
  Piotr Kulicki and Robert Trypuz

The Matter of Objects   121
  Henry Laycock

Burge's Contextual Theory of Truth and the Super-Liar
Paradox   141
  Matt Leonard

Transparent Intensional Logic. A Challenge   153
  Pavel Materna

Sceptical and Credulous Approach to Deductive Argumentation 181
    Svatopluk Nevrkla

Logical Form and Reflective Equilibrium 191
    Jaroslav Peregrin and Vladimír Svoboda

Implications as Rules in Dialogical Semantics 211
    Thomas Piecha and Peter Schroeder-Heister

Archetypal Rules and Intermediate Logics 227
    Tomasz Połacik

Semantic Paradoxes and Transparent Intensional Logic 239
    Jiří Raclavský

Unlimited Possibilities 253
    Gonçalo Santos

Boxes Are Relevant 265
    Igor Sedlár

Trying to Model Metaphor 279
    Krister Segerberg

Truth & Knowledge in Logic & Mathematics 289
    Gila Sher

A Unifying Framework for Reasoning about Normative Conflicts

Mathieu Beirlaen[*]

Abstract

First, two context-dependent desiderata are presented for devising calculi of deontic logic that can consistently accommodate normative conflicts. Conflict-tolerant deontic logics (CTDLs) can be evaluated by their treatment of the trade-off between these desiderata. Next, it is argued that CTDLs defined within the standard format for adaptive logics are particularly good at overcoming this trade-off.

Keywords: deontic logic, normative conflicts, conflict-tolerance, adaptive logic, non-monotonic logic

1 Normative conflicts

One of the many challenges in the field of deontic logic concerns the consistent accommodation of normative conflicts by formal calculi. Intuitively, a *normative conflict* occurs whenever we find ourselves in a situation in which our normative directives are inconsistent or not uniquely action-guiding in the sense that we are permitted or even obliged to do something that is forbidden. We may for instance be permitted to break a promise in view of a more binding obligation to rescue someone in need.

Normative conflicts also occur in e.g. jurisprudence (cfr. Alchourrón & Bulygin, 1971) and theoretical ethics. As a classic example of a

---

[*] I am very grateful to the organizers and participants of the *Logica 2011* conference. I also wish to thank Christian Straßer for many valuable comments and suggestions.

normative conflict in the latter context, consider one of the many formulations of the so-called trolley problem: a trolley is headed toward five people walking on the track, and its conductor has fainted. An agent is standing next to a switch, which she can throw, that will turn the trolley onto a parallel side track, thereby preventing it from killing the five people. However, there is a man standing on the side track with his back turned. If the agent throws the switch, she will kill this man. Should the agent throw the switch, or shouldn't she interfere?

2 Accommodating normative conflicts

Let the operators 'O' and 'P' represent obligations, resp. permissions. Using these operators, we can distinguish between various types of normative conflicts (Beirlaen, Straßer, & Meheus, in press):

- OO-conflicts between two or more obligations, e.g. $Op \land O\neg p$,

- OP-conflicts between an obligation and a permission, e.g. $Op \land P\neg p$,

- contradictory obligations [permissions]: conflicts between an obligation [permission] and its negation, e.g. $Op \land \neg Op$, $Pq \land \neg Pq$.

In turn, these various types can be combined into more complex conflicts, e.g. $Op \land Oq \land O\neg(p \land q)$, $O(p \land \neg q) \land (O\neg p \lor Pq)$, or $P(p \land q) \land (Pp \supset O\neg q)$. Depending on the properties of the logic in question, one or more types of normative conflicts may imply or be equivalent to a conflict of another type. For instance, in systems in which—for any well-formed formula A—$P\neg A \equiv \neg OA$, an OP-conflict $Op \land P\neg p$ is equivalent to the pair of contradictory obligations $Op \land \neg Op$.

In Standard Deontic Logic (**SDL**),[1] normative conflicts cause explosion: **SDL** cannot consistently accommodate OO-conflicts, OP-conflicts, or contradictory obligations [permissions]. The modal logician can try to overcome this limitation in one of two ways.

A first way of trying to make deontic logic more conflict-tolerant consists of enriching the language of **SDL**. This can be done by adding

---

[1]**SDL** is obtained by adding to the basic normal modal logic **K** the axiom schema $\Box A \supset \Diamond A$ (and by subsequently replacing instances of the alethic modal operator $\Box$ [$\Diamond$] by the deontic operator O [P]). This logic is also known as **KD** or simply **D**.

sub- and/or superscripts to the deontic operators for indicating the authorities, normative standards and/or interest groups in view of which the conflicting norms hold (e.g. Kooi & Tamminga, 2008). For instance, in the trolley example above we could use different superscripts for indicating that the obligation to throw the switch holds in view of the 'utilitarian' conviction that saving five lives is preferable to losing one life, whereas the obligation not to throw the switch holds in view of the 'deontological' conviction that refrainment is preferable to actively and consciously killing someone.

Another way to enrich the language of **SDL** is to introduce a preference ordering on our obligations and permissions in order to resolve conflicts between norms of different hierarchies in the order (e.g. Hansson, 2001). Doing this would allow us to model situations in which more binding obligations or permissions override less binding ones. Yet another extension of **SDL** consists in making its deontic operators dyadic in order to properly express under which conditions our obligations and permissions hold true (for a good oversight, see Åqvist, 2002).

These extensions are very successful in increasing the expressive power of **SDL**. Furthermore, they effectively allow us to consistently model conflicts between norms of different hierarchies, norms issued by different authorities, norms arising from different normative standards, norms that hold in different circumstances, etc. However, they fall short of tolerating so-called *symmetrical* normative conflicts (Sinnott-Armstrong, 1988). These are normative conflicts of the same preference, arising from one and the same authority and normative standard, that hold in view of one and the same interest group in the same circumstances.

Imagine, for instance, a situation where two identical twins are drowning and the situation is such that some agent, say Ann, can save either one of them, but she cannot save both (because, for instance, Ann is not a very good swimmer and there is not enough time to save both). Morally, Ann ought to save the first twin (Ot_1) and she ought to save the second twin (Ot_2). However, it is impossible for her to save both twins ($\neg \Diamond(t_1 \wedge t_2)$) (Gowans, 1987, p. 192). Following Lou Goble (2005), we can represent Ann's dilemma in purely deontic terms by means of the formulas $Ot_1 \wedge O\neg t_1$ and $Ot_2 \wedge O\neg t_2$.[2] Given our assumptions, these OO-conflicts are perfectly symmetrical and will still cause explosion in any of the proposed enrichments of **SDL**.

A second way of allowing for the consistent possibility of normative conflicts in deontic logic consists in weakening **SDL**. Instead of making **SDL** more expressive, this approach proceeds by rejecting or restricting certain **SDL**-valid inferences so as to avoid explosion in view of normative conflicts. The aim is not to extend the language of **SDL**, but rather to avoid absurdities from following from a normative conflict. Where NC is an OO- or OP-conflict or a pair of contradictory obligations or permissions, and where **L** is some logic, we also want to avoid things like (Goble, 2005):

$$NC \vdash_\mathbf{L} OB \qquad \text{(DEX-1)}$$
$$NC \vdash_\mathbf{L} PB \supset OB \qquad \text{(DEX-2)}$$
$$NC \vdash_\mathbf{L} OB \vee O\neg B \qquad \text{(DEX-3)}$$

Likewise for premise sets containing more complex or combined normative conflicts. By weakening **SDL** to a system that invalidates principles like (DEX-1), (DEX-2), and (DEX-3), we can arrive at a logic in which we can sensibly accommodate even the symmetrical cases discussed above. For this reason and for the reason that we might simply be constrained by a not very expressive formal language, I focus on this latter approach from now on.

Where $i \in \{1, 2, 3\}$, the variants of (DEX-i) where NC is a OO-conflict (resp. OP-conflict, pair of contradictory norms) are in the remainder denoted by (OO-DEX-i) (resp. (OP-DEX-i), ($\bot$-DEX-i)). Henceforth, logics devised in order to consistently accommodate one or more types of normative conflicts are called *conflict-tolerant deontic logics* (CTDLs).

3 Deontic pluralism

The aim of this paper is not to present and defend one particular conflict-tolerant deontic logic that allows for the consistent possibility

---

[2] Goble arrives at this representation via the principle $\neg\Diamond(A \wedge \neg B) \supset (OA \supset OB)$, by means of which the obligations $O\neg t_1$ and $O\neg t_2$ follow from Ot_1, Ot_2, and $\neg\Diamond(t_1 \wedge t_2)$.

of *all* types of normative conflicts. I believe instead that the adequacy of a given CTDL is a context-dependent matter: both its rules of inference and its degree of conflict-tolerance depend on the concrete application of the logic. Let me illustrate this claim by means of three examples, each of which is situated in a different 'deontic' context.

(1) As a first example, consider a *moral* context. In discussions on moral dilemmas, philosophers have typically focussed on conflicting obligations. Moral dilemmas are conceived as situations in which an agent ought to adopt each of two or more alternatives which are equally compelling from a moral point of view, and in which the agent cannot do both (or all) of the actions (e.g. Sinnott-Armstrong, 1988). In view of the discussion in Section 2, moral dilemmas can be formalized as OO-conflicts (cfr. footnote 2).

Since there is nothing particularly 'dilemmatic' about an OP-conflict (here, the agent can still safely fulfill all of her moral requirements, i.e. all of her obligations), then—assuming that a rational agent facing a moral dilemma is not facing any contradictory obligations or permissions—a CTDL that allows for the consistent possibility of OO-conflicts is sufficiently conflict-tolerant for dealing with moral dilemmas.

How, then should **SDL** be weakened in this context? One suggestion is to reject or restrict the aggregation schema (AND):

$$(OA \wedge OB) \supset O(A \wedge B) \qquad \text{(AND)}$$

In the moral context, (AND) was disputed (amongst others) by Bernard Williams, who argued that an agent facing an OO-conflict thinks that she should fulfill each of the conflicting obligations, but does not think that she should fulfill both (Williams, 1965).

(2) Next, consider the context of *normative systems*. In this context, formulas of the form OA [PA] are read as "there exists a norm to the effect that A is mandatory [permitted]". Even though jurists have created a number of principles in order to resolve legal conflicts, normative systems often contain irresolvable conflicts between norms. These conflicts can be formalized as OO- or OP-conflicts (e.g. Alchourrón, 1969; Alchourrón & Bulygin, 1971).

In the context of normative systems, formulas of the form OA or PA abbreviate statements *about* norms. For instance, a formula $\neg Op$ [$\neg Pp$] denotes the absence of a norm to the effect that p is mandatory [permitted].³ Whereas a normative system may very well contain both a norm to the effect that p is mandatory as well as a norm to the effect that $\neg p$ is mandatory or permitted, it is less clear how such a system could both contain and not contain a norm to the effect that p is mandatory or permitted.⁴ Thus it is reasonable to construct a logic of normative systems that takes into account the possibility of OO- and OP-conflicts, but not the possibility of contradictory norms.

Due to the possibility of OP-conflicts and the specific interpretation of the deontic operators in this context, a concrete CTDL for normative systems should invalidate the interdefinability schema (DfP) (Wright, 1963):

$$\mathsf{P}A \equiv \neg \mathsf{O} \neg A \tag{DfP}$$

(3) As a third and final illustration, consider the logic of commands. In this setting, the O- and P-operators are interdefinable, i.e. (DfP) is valid in this context (see e.g. Wright, 1963). Since it is possible for a (confused) authority to assert that p is obligatory, and also that $\neg p$ is obligatory or permitted, OO- and OP-conflicts should be tolerated. Moreover, we should allow for contradictory obligations and permissions in this setting, since a formula $\mathsf{O}A \wedge \mathsf{P}\neg A$ is equivalent to $\mathsf{O}A \wedge \neg \mathsf{O}A$ and $\neg \mathsf{P}\neg A \wedge \mathsf{P}\neg A$ in view of (DfP). In (Beirlaen et al., in press), a CTDL is presented that is *fully* conflict-tolerant in the sense that it invalidates all of (DEX-1), (DEX-2), and (DEX-3) for all types of normative conflicts presented above. This logic weakens **SDL** by turning its classical negation into a paraconsistent one, thus invalidating the *Ex Contradictione Quodlibet* schema:

$$(A \wedge \neg A) \supset B \tag{ECQ}$$

One need not agree with all the details in illustrations (1)–(3) in order to be convinced by the main argument, namely that different normative contexts require different CTDLs. From the illustrations, it is also clear that the degree of conflict-tolerance of a given CTDL, i.e. the variety of types of normative conflicts that the CTDL should consistently allow for, is also context-dependent.

---

[3] Formulas of the form "PA" are interpreted here as *strong* or *positive* permissions, in accordance with their interpretation in (Alchourrón, 1969).

[4] Exceptions can be made, for instance, when one of two parties argues that system S *does* contain a norm to the effect that A is permitted, whereas the other argues that S *doesn't* contain such a norm. However, such a context is different from the one discussed here.

Given these insights, we can formulate a first of two desiderata for CTDLs:

Desideratum 1 *Given the normative context to which it is applied, a CTDL should be sufficiently conflict-tolerant.*

More formally, this desideratum boils down to the demand that, depending on the types of conflicts we want our logic **L** to be able to accommodate, **L** should invalidate principles like (DEX-1)–(DEX-3) for these types of conflicts. If, for instance, we want to devise a logic **L** for dealing with moral dilemmas as in illustration (1), then—according to Desideratum 1—**L** must invalidate (at least) (OO-DEX-1), (OO-DEX-2), and (OO-DEX-3).

4 CTDLs and inferential power

By weakening **SDL** in order to make it more conflict-tolerant, we run the risk of losing inferences that are intuitively valid. Suppose, for instance, that we want to devise a CTDL for dealing with moral dilemmas, as in illustration (1) in Section 3. Suppose further that we follow Williams' suggestion and reject the (AND)-schema. Then consider the following example from (Horty, 1994, p. 39): suppose that some agent should either fight in the army or perform alternative service to his country ($O(f \vee s)$). As a pacifist, this agent is opposed to warfare, so he ought not to fight in the army ($O\neg f$). Then he can consistently satisfy all of his obligations by performing alternative service to his country. However, although Os is **SDL**-derivable from $O(f \vee s)$ and $O\neg f$, we can no longer make this inference if we give up (AND). This led Horty to the conclusion that "apparently, what is needed is some degree of agglomeration [aggregation], but not too much; and the problem of formulating a principle allowing for exactly the right amount of agglomeration [aggregation] raises delicate issues that have generally been ignored in the literature" (Horty, 2003, p. 580).

Horty's argument boils down to the need for validating at least *some* instances of the Deontic Disjunctive Syllogism schema:

$$(O(A \vee B) \wedge O\neg A) \supset OB \qquad \text{(DDS)}$$

(DDS) is lost in its entirety if we reject all instances of (AND). In general, Horty's example illustrates the need for the following (second) desideratum for CTDLs:

Desideratum 2 *Given the normative context, a CTDL should be strong enough to account for all intuitively valid inferences.*

The main problem then in devising adequate CTDLs consists in finding the right equilibrium between Desideratum 1 and Desideratum 2: we want a logic that is sufficiently *weak* in order to meet the first, and sufficiently *strong* to meet the second desideratum.

5 The adaptive logics framework

In the remainder of this paper, I defend the 'adaptive' approach for devising and evaluating CTDLs. Logics developed within the adaptive logics framework are well-suited for modeling non-monotonic reasoning. Like most human inferencing, our normative reasoning is non-monotonic. We may, for instance, withdraw a conclusion Op drawn from two premises $O(p \vee q)$ and $O \neg q$ in view of the new information $O \neg p$. In order for a logic $\mathbf{L}$ to model such reasoning processes, it needs to be non-monotonic: $O(p \vee q), O \neg q \vdash_{\mathbf{L}} Op$, yet $O(p \vee q), O \neg q, O \neg p \nvdash_{\mathbf{L}} Op$.

Most adaptive CTDLs (ACTDLs) devised so far are defined within the *standard format* for adaptive logics from (Batens, 2007). An ACTDL in this standard framework is characterized as a triple, consisting of:

1. A lower limit logic **LLL**: a reflexive, transitive, monotonic and compact CTDL that contains classical logic and has a characteristic semantics.

2. A set of abnormalities Ω: a set of **LLL**-contingent formulas, characterized by a logical form $\mathcal{F}$; or a union of such sets.

3. An adaptive strategy: Reliability or Minimal Abnormality.

The lower limit logic is the stable part of the adaptive logic (AL); anything that follows from the premises by **LLL** will never be revoked. In the case of CTDLs, we want this 'base' logic to be sufficiently conflict-tolerant given its context of application, i.e. the lower limit logic of an ACTDL has to meet Desideratum 1.

Typically, an AL enables one to derive, for most premise sets, some additional consequences on top of those that are **LLL**-derivable.

These supplementary consequences are obtained by interpreting a premise set "as normally as possible", or, equivalently, by supposing abnormalities to be false "as much as possible". For sensible ACTDLs, the set Ω is defined in such a way that, for every type of normative conflict that we want to be able to accommodate, a member of Ω is **LLL**-derivable from a normative conflict. A concrete illustration of this idea follows in Section 6.

The formal disambiguation of the phrases "as normally as possible" and "as much as possible" in the previous paragraph is relative to the adaptive strategy used. The two strategies currently defined within the standard framework are Reliability and Minimal Abnormality. Generally, the Minimal Abnormality strategy allows one to derive some extra consequences on top of those derivable by means of the Reliability strategy.

ALs defined within the standard format are well-behaved syntactically and semantically. Syntactically, adaptive proofs extend a Fitch-style proof theory (Fitch, 1952), which is illustrated informally in Section 6. Semantically, ALs have a Shoham-style preferential model semantics (Shoham, 1987). A detailed account of the semantics and proof theory for ALs in the standard format can be found in (Batens, 2007).

6 Illustration: the logic DP^r

The logic **DP^r** from (Beirlaen et al., in press) aims to meet the demands described in illustration (3) of Section 3. Its lower limit logic **DP** is a fully conflict-tolerant paraconsistent CTDL:[5] it invalidates (DEX-1)–(DEX-3) for any type of normative conflict. Its set of abnormalities is the set of well-formed formulas of the form $A \wedge \neg A$, where '$\neg$' is a paraconsistent negation connective, and where A is either an atomic proposition or a formula of the form OB, where B is a well-formed formula of classical propositional logic. In view of the construction of **DP**, an abnormality is derivable from any normative conflict: $OA \wedge O\neg A \vdash_{\mathbf{DP}} O\neg A \wedge \neg O\neg A$, $OA \wedge P\neg A \vdash_{\mathbf{DP}} OA \wedge \neg OA$, and $PA \wedge \neg PA \vdash_{\mathbf{DP}} O\neg A \wedge \neg O\neg A$. The strategy employed by **DP^r** is Reliability (hence the superscript r).

Due to its invalidation of all instances of (DEX-1)–(DEX3) for all

---
[5] A logic is *paraconsistent* if it invalidates the (ECQ) schema.

types of normative conflicts, **DP** meets Desideratum 1. However, it does not meet Desideratum 2. It is, for instance, too weak in order to account for Horty's example from Section 4: $O(f \vee s), O\neg f \nvdash_{\mathbf{DP}} Os$. This problem is solved by **DP^r**. Let $\Gamma = \{O(f \vee s), O\neg f\}$. We start a **DP^r**-proof from Γ by introducing the premises. This can be done in an adaptive proof via the premise introduction rule PREM:

    1   $O(f \vee s)$ PREM $\emptyset$
    2   $O\neg f$ PREM $\emptyset$

The last column in the proof is called the *condition*. For lines at which premises are introduced, the condition is always empty. Due to the paraconsistency of "$\neg$", the Disjunctive Syllogism rule is **DP**-invalid. So is its deontic variant (DDS), which would allow us to derive Os from $O(f \vee s)$ and $O\neg f$. However, the weaker formula $Os \vee (O\neg f \wedge \neg O\neg f)$ *is* **DP**-derivable from Γ. We can add this formula to the proof by using the *unconditional rule* RU. This rule allows us to derive—for any adaptive logic in the standard format—all formulas that follow from the premises by the lower limit logic. In the condition column of a line at which RU is applied, we find the union of the conditions found at the lines used in this derivation. The formula at line 3 below is derived on the empty condition because it relies on the formulas at lines 1 and 2, the conditions of which are empty too.

    3   $Os \vee (O\neg f \wedge \neg O\neg f)$ 1,2; RU $\emptyset$

Note that the second disjunct of the formula derived at line 3 is a member of Ω. At a line in an adaptive proof, we can move to the condition those members of Ω which were derived in disjunction with some other formula. This is realized by an application of the *conditional rule* RC:

    4   Os 3; RC $\{O\neg f \wedge \neg O\neg f\}$

Informally, moving an abnormality to the condition column corresponds to making the assumption that this abnormality is false. Thus, at line 4 we have derived the formula Os on the assumption that $O\neg f \wedge \neg O\neg f$ is false.

If it would later turn out that $O\neg f \wedge \neg O\neg f$ were true after all, then the lines in the proof at which we assumed this formula to be false would become *marked* in the proof. In fact, for the Reliability strategy lines become marked as soon as an element of their condition occurs in a disjunction of abnormalities that follows from the premise set by means of the lower limit logic.

Since no such disjunction containing the formula $O\neg f \wedge \neg O\neg f$ is **DP**-derivable from Γ, line 4 remains unmarked in any extension of this proof. By the criterion for final derivability in an adaptive proof, it follows that Os is $\mathbf{DP^r}$-derivable from Γ: $\Gamma \vdash_{\mathbf{DP^r}} Os$.[6]

As opposed to its lower limit logic, $\mathbf{DP^r}$ meets Desideratum 2. In fact, for all **SDL**-consistent premise sets Γ, $\Gamma \vdash_{\mathbf{DP^r}} A$ iff $\Gamma \vdash_{\mathbf{SDL}} A$. In adaptive terminology: **SDL** is the *upper limit logic* of $\mathbf{DP^r}$.

For reasons of space, I cannot here provide more details about the workings of $\mathbf{DP^r}$. For a full formal account of the proof theory of $\mathbf{DP^r}$ (including its marking definition and the definition of final derivability in a $\mathbf{DP^r}$-proof), for its semantics, and for more illustrations for this logic, see (Beirlaen et al., in press).

7 Conclusion and outlook

The logic $\mathbf{DP^r}$ is but one of a large family of adaptive CTDLs. Other adaptive CTDLs can be found in (e.g. Goble, n.d.; Meheus, Beirlaen, & Van De Putte, 2010; Straßer, 2010; Straßer, Beirlaen, & Meheus, in press).

Assuming a pluralistic (context-dependent) attitude with respect to CTDLs, adaptive logics provide a unifying framework for studying various CTDLs in different normative contexts. The list of normative contexts and respective CTDLs hinted at in this paper is not meant to be exhaustive. Different options are available for the deontic logician in devising (A)CTDLs.

Of course, the adaptive logics framework itself only represents one of the many options open for devising CTDLs. Its advantages include

---

[6]From the informal statement of the marking criterion for Reliability, it is clear that line 4 would become marked in a proof from $\Gamma \cup \{\neg O\neg f\}$ as soon as the abnormality $O\neg f \wedge \neg O\neg f$ is derived in the proof (since $\Gamma \cup \{\neg O\neg f\} \vdash_{\mathbf{DP}} O\neg f \wedge \neg O\neg f$, the latter formula is obtainable via RU on the empty condition). In view of the marking definition for Reliability, and the criterion for final derivability in a $\mathbf{DP^r}$-proof, it follows that $\Gamma \cup \{\neg O\neg f\} \nvdash_{\mathbf{DP^r}} Os$. This illustrates the non-monotonicity of $\mathbf{DP^r}$.

a very intuitive treatment of the trade-off between Desiderata 1 and 2, and a well-behaved semantics and defeasible proof theory. Moreover, the framework itself is flexible in the sense that it allows for various extensions and enrichments in the sense discussed in Section 2 (for an example, see Beirlaen & Straßer, 2011; Van De Putte & Straßer, in press). Furthermore, ACTDLs defined within the standard format come with a meta-theory which automatically guarantees properties like smoothness, fixed point, soundness and completeness (see Batens, 2007 for proofs and a more detailed list of meta-theoretical properties).

One drawback of this framework is that adaptive logics tend to be computationally complex. But then again, so is human (normative) reasoning (Batens, de Clercq, Verdée, & Meheus, 2009).

References

Alchourrón, C. E. (1969). Logic of norms and logic of normative propositions. *Logique & Analyse, 47*, 242–268.

Alchourrón, C. E., & Bulygin, E. (1971). *Normative systems.* Springer-Verlag, Wien/New York.

Åqvist, L. (2002). Deontic logic. In D. Gabbay & F. Guenthner (Eds.), *Handbook of philosophical logic (2nd edition)* (Vol. 8, pp. 147–264). Kluwer Academic Publishers.

Batens, D. (2007). A universal logic approach to adaptive logics. *Logica Universalis, 1*, 221–242.

Batens, D., de Clercq, K., Verdée, P., & Meheus, J. (2009). Yes fellows, most human reasoning is complex. *Synthese, 166*(1), 113–131.

Beirlaen, M., & Straßer, C. (2011). A paraconsistent multi-agent framework for dealing with normative conflicts. In J. Leite, P. Torrini, T. Ågotnes, G. Boella, & L. van der Torre (Eds.), *Computational logic in multi-agent systems. Proceedings of the 12th international workshop CLIMA XII* (Vol. 6814, pp. 312–329). Springer-Verlag.

Beirlaen, M., Straßer, C., & Meheus, J. (in press). An inconsistency-adaptive deontic logic for normative conflicts. *Journal of Philosophical Logic.* (In print. DOI: 10.1007/s10992-011-9221-3)

Fitch, F. B. (1952). *Symbolic logic: An introduction.* Ronald Press Co.

Goble, L. (n.d.). Notes on adaptive deontic logics for normative conflicts. *Unpublished paper*.

Goble, L. (2005). A logic for deontic dilemmas. *Journal of Applied Logic*, *3*, 461–483.

Gowans, C. W. (Ed.). (1987). *Moral dilemmas*. Oxford University Press.

Hansson, S. O. (2001). *The structure of values and norms*. Cambridge University Press.

Horty, J. F. (1994). Moral dilemmas and nonmonotonic logic. *Journal of Philosophical Logic*, *23*(1), 35–66.

Horty, J. F. (2003). Reasoning with moral conflicts. *Noûs*, *37*, 557–605.

Kooi, B., & Tamminga, A. (2008). Moral conflicts between groups of agents. *Journal of Philosophical Logic*, *37*, 1–21.

Meheus, J., Beirlaen, M., & Van De Putte, F. (2010). Avoiding deontic explosion by contextually restricting aggregation. In G. Governatori & G. Sartor (Eds.), *Deon (10th international conference on deontic logic in computer science)* (Vol. 6181, pp. 148–165). Springer.

Shoham, Y. (1987). A semantical approach to nonmonotonic logics. In M. L. Ginsberg (Ed.), *Readings in nonmonotonic reasoning* (pp. 227–250). Morgan Kaufmann Publishers.

Sinnott-Armstrong, W. (1988). *Moral dilemmas*. Basil Blackwell, Oxford/New York.

Straßer, C. (2010). An adaptive logic framework for conditional obligations and deontic dilemmas. *Logic and Logical Philosophy*, *19*(1–2), 95–128.

Straßer, C., Beirlaen, M., & Meheus, J. (in press). Tolerating deontic conflicts by adaptively restricting inheritance. *Logique & Analyse*. (To appear)

Van De Putte, F., & Straßer, C. (in press). A logic for prioritized normative reasoning. *Journal of Logic and Computation*. (In print. DOI: 10.1093/logcom/exs008)

Williams, B. (1965). Ethical consistency. *Proceedings of the Aristotelian Society (Supplementary Volumes)*, *39*, 103–124.

Wright, G. H. von. (1963). *Norm and action. A logical enquiry*. Routledge and Kegan Paul, London.

Mathieu Beirlaen
Centre for Logic and Philosophy of Science, Ghent University
Blandijnberg 2, B-9000 Gent, Belgium
e-mail: `Mathieu.Beirlaen@UGent.be`

Non-Normal Worlds and Representation

Francesco Berto*

Abstract

World semantics for relevant logics include so-called *non-normal* or *impossible* worlds providing model-theoretic counterexamples to such irrelevant entailments as $(A \land \neg A) \to B$, $A \to (B \lor \neg B)$, or $A \to (B \to B)$. Some well-known views interpret non-normal worlds as information states. If so, they can plausibly model our ability of *conceiving* or *representing* logical impossibilities. The phenomenon is explored by combining a formal setting with philosophical discussion. I take Priest's basic relevant logic N_4 and extend it, on the syntactic side, with a representation operator, ®, and on the semantic side, with particularly anarchic non-normal worlds. This combination easily invalidates unwelcome "logical omniscience" inferences of standard epistemic logic, such as belief-consistency and closure under entailment. Some open questions are then raised on the best strategies to regiment ® in order to express more vertebrate kinds of conceivability.

Keywords: impossible worlds, relevant logic, logical omniscience, conceivability

1 Overview

Relevant logics are perhaps the most developed among paraconsistent logics, these being logical systems rejecting the principle *ex contradictione quodlibet* (ECQ), according to which a contradiction entails everything (in 'object language' version, $(A \land \neg A) \to B$). Arguably, the most discussed kinds of formal semantics for relevant logics are

---

*Thanks to the audience at the Logica 2011 conference for comments and helpful feedback.

world semantics. As specialists know, these include so-called *non-normal* or *impossible* worlds, often thought of as situations where the truth conditions of logical operators are different.

Non-normal worlds are crucial for providing model-theoretic counterexamples to ECQ as well as to other irrelevant entailments, such as $A \to (B \vee \neg B)$ and $A \to (B \to B)$.[1] They can thus help in modeling our capacity of reasoning non-trivially also in the face of inconsistent information. And such a capacity is widely attested, thus providing counterexamples to ECQ. For an often mentioned case: Bohr's atomic theory includes both the assumption that energy has the form of quanta, that is, discrete packs, and Maxwell's usual electromagnetic equations, which are inconsistent with that assumption.[2] Nevertheless, Bohr provided quite a successful theory. More importantly for our purposes: he did not infer arbitrary conclusions from his inconsistent assumptions—for instance, that electrons have the same electric charge as protons.

The main philosophical issue concerning world semantics for relevant logics has traditionally been the one of the intuitive reading of its worlds, and of the relations and operations defined on them. Some well-known views interpret these precisely as information states, or conduits thereof (see e.g. Mares, 2004). Given such an epistemically-driven reading, non-normal worlds may model our ability of *conceiving* or *representing* inconsistencies and broadly logical impossibilities. This is tightly connected to our aforementioned capacity of reasoning efficiently in inconsistent informational circumstances—if not a precondition of it. As Bohr knew he was making incompatible assumptions in his theory, for instance, he was arguably able to conceive those

---

[1] Intuitively, a premise or conditional antecedent is irrelevant within an inference or a conditional, if it is of no utility in getting to the conclusion, or in grounding the consequent. The research program of relevant logic is based on the positive view that the intuition of relevance can be given formal substance, together with the negative view that classical logic legitimates irrelevant inferences—on the ground, for instance, of its admitting logically valid conditionals with no content connection between antecedent and consequent. At least part of the formal substance to the idea of relevance as content-connection is provided by the so-called Variable Sharing Property (VSP), also called weak or necessary condition of relevance. As far as the conditional goes, this states that if $A \to B$ is logically valid, then A and B must share some sentential variable. On this ground, ECQ and the two aforementioned formulas count as fallacies of relevance, not passing the VSP test. For a short and accessible introduction to relevant logic, see (Mares, 2004).

[2] For an account of this story, see (Brown, 1993).

inconsistent suppositions as holding together. This did not lead him astray, though.

Supposing the non-normal worlds of relevant semantics are essentially realizations of *intentional* states, such as conceiving or representing,[3] this paper explores the phenomenon by combining a formal setting with philosophical discussion.

I proceed as follows: in section 2, I introduce the syntax of a first-order intensional language L and, in section 3, I present a model-theoretic semantics for it, which draws upon the relevant logic N_4 proposed in chapter 9 of (Priest, 2001). This combines techniques of many-valued and modal logics, including locally inconsistent and incomplete non-normal worlds, but has standard definitions of logical consequence and validity. Despite being simpler than the mainstream world semantics for relevant logics, the N_4 setting allows to model all the fetures of relevant systems that are significant for our purposes in a friendly formal setting. The language includes a representation operator, whose role is to capture our capacity of representing or conceiving inconsistencies and logical impossibilities.

Section 4 provides a brief discussion of the distinction, embedded in the model, between two kinds of non-normal worlds, displaying different degrees of logical lawlessness and labeled, for reasons to be explained, as *extensionally* and *intensionally* impossible worlds.

In section 5, it is shown that the semantics makes of L's conditional a fully relevant (albeit weak) one, invalidating the fallacies of relevance and, in particular, ECQ.

Section 6 explains how the representation operator invalidates (the formulations in terms of it of) typical unwelcome inferences of epistemic logic gathered under the rubric of "logical omniscience", such as belief-consistency and closure under entailment. It is well known that logical omniscience phenomena make for highly idealized epistemic notions, not mirroring the actual condition of human beings as finite, fallible, and occasionally inconsistent cognitive agents. If we can conceive inconsistencies and impossibilities, non-normal worlds are natural candidates to model this condition: the content of a rep-

---

[3] I employ these two terms as generics for a range of broadly cognitive human activities, all involving the depiction of scenarios, situations, or circumstances, which count as their contents. I take a dim view on such intentional phenomena, and leave their serious investigation to philosophers of mind, cognitive scientists, or neuroscientists.

resentational state is the set of worlds that make the representation true, that is, where things are as they are conceived or represented to be. This may include non-normal worlds where those inferences fail.

Finally, some open questions are raised in section 7, as to the best strategies to regiment the representation operator in order for it to express specific and more vertebrate kinds of conceivability; these should intuitively be closed under some (albeit weaker-than-classical) logical consequence relation and, more importantly, allow for *ceteris paribus* import of information from actuality.

2 Syntax of L

L consists of a fully standard first-order vocabulary with individual variables x, y, z (and, if more are needed, indexed ones, x_1, ..., x_n); individual constants: m, n, o (if more are needed, m_1, ..., m_n); n-place predicates: F, G, H (F_1, ..., F_n); the usual connectives, negation $\neg$, conjunction $\wedge$, disjunction $\vee$, the conditional $\rightarrow$; the two quantifiers, $\forall$ and $\exists$; the two standard alethic modal operators for necessity $\Box$ and possibility $\Diamond$; a unary sentential operator Ⓡ; round brackets as auxiliary symbols. Individual constants and variables are singular terms. If t_1, ..., t_n are singular terms and P is any n-place predicate, Pt_1, ..., t_n is an atomic formula. If A and B are formulas, $\neg A$, $(A \wedge B)$, $(A \vee B)$, $(A \rightarrow B)$, $\Box A$, $\Diamond A$ and ⓇA are; outermost brackets are normally omitted in formulas. If A is a formula and x is a variable, then $\forall x A$ and $\exists x A$ are formulas, closed and open formulas having their standard definitions.

The only piece of notational novelty is Ⓡ, which I shall call the *representation* operator. The intuitive reading of 'ⓇA' will be "It is represented that A", or "It is conceived that A".

3 Semantics for L

The semantics for L is largely down to Priest's work in non-standard intensional logic (see Priest, 2001, 2005), with a few modifications. An interpretation is an ordered septuple $\langle P, I, E, @, R, D, v \rangle$, the intuitive reading of whose members is as follows. P is the familiar set of possible worlds; I and E are two sets of non-normal or impossible worlds of two kinds, the *intensionally* and *extensionally* impossible ones respectively (what this means, we will see soon); P, I and E are disjoint, W = P ∪ I ∪ E is the totality of worlds *simpliciter*. @ is the obtaining world (or, better, its foster in the formalism). I assume, for prudence, that @ ∈ P. $R \subseteq W \times W$; if $\langle w_1, w_2 \rangle \in R$ ($w_1, w_2 \in W$), I write this as '$w_1 \, R \, w_2$' and claim that w_2 is *representationally accessible* (R-accessible), from w_1 (what this means, we will also see soon). D is a non-empty set of objects. v is a function assigning denotations to the descriptive constant symbols of L, as follows:

If c is an individual constant, $v(c) \in D$.

If P is an n-place predicate and $w \in W$, $v(P, w)$ is a pair: $\langle v^+(P, w), v^-(P, w) \rangle$, with $v^+(P, w) \subseteq D^n$, $v^-(P, w) \subseteq D^n$.

$D^n = \{ \langle d_1, \ldots, d_n \rangle \mid d_1, \ldots, d_n \in D \}$, and $\langle d \rangle$ is stipulated to be just d, so D^1 is D. To each pair of n-place predicate P and world w, v assigns a (positive) extension $v^+(P, w)$ and an anti-extension or negative extension, $v^-(P, w)$. The extension of P at w is to be thought of as the set of (n-tuples of) things of which P is true there, the anti-extension as the set of (n-tuples of) things of which P is false there. Such double extensions are to model inconsistencies—things being both true and false (truth value gluts; or also, neither true nor false—truth value gaps). On the other hand, one may sensibly want truth and falsity to be exclusive and exhaustive at possible worlds (this is part of what makes them possible, after all). We can recover the classical setting by imposing the following double clause—let us call it the Classicality Condition:

(CC) If $w \in P$, for any n-ary predicate P: $v^+(P, w) \cap v^-(P, w) = \emptyset$, $v^+(P, w) \cup v^-(P, w) = D^n$.

At possible worlds, extensions and anti-extensions are exclusive and exhaustive. We need the usual assignments of denotations to variables. If a is an assignment (a map from the variables to D), then v_a is the suitably parameterized denotation function, so that we have denotations for all singular terms:

If c is an individual constant, $v_a(c) = v(c)$.

If x is a variable, $v_a(x) = a(x)$.

Let us read '$w \Vdash_a^+ A$' as "A is true at world w (with respect to assignment a)", and '$w \Vdash_a^- A$' as "A is false at world w (with respect to assignment a)" (and an interpretation, but I will omit to mention it when no confusion arises). The truth and falsity conditions for atomic formulas are:

$w \Vdash_a^+ Pt_1 \ldots t_n$ iff $\langle v_a(t_1), \ldots, v_a(t_n) \rangle \in v^+(P, w)$
$w \Vdash_a^- Pt_1 \ldots t_n$ iff $\langle v_a(t_1), \ldots, v_a(t_n) \rangle \in v^-(P, w)$

The extensional vocabulary has straightforward clauses at all $w \in P \cup I$:

$w \Vdash_a^+ \neg A$ iff $w \Vdash_a^- A$
$w \Vdash_a^- \neg A$ iff $w \Vdash_a^+ A$

$w \Vdash_a^+ A \wedge B$ iff $w \Vdash_a^+ A$ and $w \Vdash_a^+ B$
$w \Vdash_a^- A \wedge B$ iff $w \Vdash_a^- A$ or $w \Vdash_a^- B$

$w \Vdash_a^+ A \vee B$ iff $w \Vdash_a^+ A$ or $w \Vdash_a^+ B$
$w \Vdash_a^- A \vee B$ iff $w \Vdash_a^- A$ and $w \Vdash_a^- B$

$w \Vdash_a^+ \forall x A$ iff for all $d \in D$, $w \Vdash_{a(x/d)}^+ A$
$w \Vdash_a^- \forall x A$ iff for some $d \in D$, $w \Vdash_{a(x/d)}^- A$

$w \Vdash_a^+ \exists x A$ iff for some $d \in D$, $w \Vdash_{a(x/d)}^+ A$
$w \Vdash_a^- \exists x A$ iff for all $d \in D$, $w \Vdash_{a(x/d)}^- A$

'$a(x/d)$' stands for the assignment that agrees with a on all variables, except for its assigning d to x. As for the modals, we have the following for all $w \in P$:

$w \Vdash_a^+ \Box A$ iff for all $w_1 \in P$, $w_1 \Vdash_a^+ A$
$w \Vdash_a^- \Box A$ iff for some $w_1 \in P$, $w_1 \Vdash_a^- A$

$w \Vdash_a^+ \Diamond A$ iff for some $w_1 \in P$, $w_1 \Vdash_a^+ A$
$w \Vdash_a^- \Diamond A$ iff for all $w_1 \in P$, $w_1 \Vdash_a^- A$

(Unrestricted) necessity/possibility is truth at all/some *possible* world(s) (I am not making much use of the box and diamond in this work, but they can be usefully contrasted, within the model, with the behaviour of the representation operator). While we have the normal material conditional, say $A \supset B =_{df} \neg A \vee B$, our more vertebrate intensional conditional is the following. At all $w \in P$:

$w \Vdash_a^+ A \to B$ iff for all $w_1 \in P \cup I$ such that $w_1 \Vdash_a^+ A$, $w_1 \Vdash_a^+ B$
$w \Vdash_a^- A \to B$ iff for some $w_1 \in P \cup I$, $w_1 \Vdash_a^+ A$ and $w_1 \Vdash_a^- B$

So far everything works familiarly enough as far as worlds in P are concerned, the main change with respect to standard modal semantics being that truth and falsity conditions are spelt separately. But even this does not change much at possible worlds. The CC dictates that, at each $w \in P$, any predicate is either true or false of the relevant object (or n-tuple thereof), but not both. That no atomic formula is both true and false or neither true nor false entails that no formula is, as can be checked recursively. Overall, there are no truth value gluts or gaps at possible worlds.[4] In particular, for instance, if $w \in P$ then $w \Vdash_a^+ \neg A$ if and only if it is not the case that $w \Vdash_a^+ A$: at possible worlds negation works "homophonically", the classical way. And since @ $\in$ P, truth *simpliciter*, truth at the actual world, behaves in an orthodox way with respect to negation.

Things get more exciting at non-normal worlds. At points in I, v treats formulas of the form $A \to B$, $\Box A$, and $\Diamond A$ essentially as *atomic*: their truth values are not determined recursively, but directly assigned by v in an arbitrary way. At points in E, all formulas can be treated as atomic and behave arbitrarily: $A \vee B$ may turn out to be true even though both A and B are false, etc.[5] Hence the denominations for the

---

[4] A technical note which may be skipped without loss of continuity. Because of the world quantifiers in the clauses for $\to$ ranging on $P \cup I$, one needs, in fact, a couple of extra assumptions on the falsity conditions for $A \to B$ to rule out gaps and gluts from possible worlds, specifically: if $w \in P$, then $w \Vdash_a^- A \to B$ iff: (1) For some $w_1 \in P \cup I$, $w_1 \Vdash_a^+ A$ and $w_1 \Vdash_a^- B$ and it is not the case that $w \Vdash_a^+ A \to B$; (2) (For some $w_1 \in P \cup I$, $w_1 \Vdash_a^+ A$ and $w_1 \Vdash_a^- B$) or it is not the case that $w \Vdash_a^+ A \to B$. A similar proviso is needed to rule out gaps and gluts for ®, given that its clauses, which we are about to meet, also allow access to non-normal worlds.

[5] Another technical note, skippable without loss of continuity. We want the syntax of various complex formulas to be semantically neglected at non-normal worlds: this is what "treating them as atomic" amounts to. But if, for instance, conditionals $A \to B$ are simply assigned arbitrary truth-values, we may have that $Fm \to Gm$ gets a different value from $Fn \to Gn$ even though m and n happen to denote the same thing, which would interact badly with the quantifiers. Priest fixes this as follows. Each formula, A, is associated with one of the form $M[x_1, \ldots, x_n]$, called the formula's *matrix*. One obtains the matrix of A by replacing each occurrence in it of a free term (either an individual constant, or a variable free in A), from left to right, with a distinct variable $x_1, \ldots, x_n$, in this order, these being indexed as the least variables greater than all the variables bound in A in some canonical ordering. One gets back the original formula from its matrix

two kinds of worlds: at intensionally impossible worlds, only the conditional and the modals are anarchic; at the extensionally impossible ones, also the extensional vocabulary behaves arbitrarily.[6]

The idea of having complex formulas behave as atomic at some worlds comes from the classic (Rantala, 1982), where non-normal worlds were introduced to make logical omniscience fail for epistemic operators. I use non-normal worlds for similar, but more general, purposes. Such worlds are to be accessible via the binary R when the truth conditions for ⓡ are at issue. At $w \in P$:

$w \Vdash_a^+$ ⓡA iff for all w_1 such that $w\,R\,w_1$, $w_1 \Vdash_a^+ A$
$w \Vdash_a^-$ ⓡA iff for some w_1 such that $w\,R\,w_1$, $w_1 \Vdash_a^- A$

The semantics for ⓡ is similar to the ordinary binary accessibility semantics for the standard modal operators. '$w\,R\,w_1$' ("w_1 is R-accessible from w"), should be read as the claim that, at w_1, things are as they are conceived or represented to be at w. So it is represented that A (at w) just in case A is true at all w_1 where things are as they are represented to be. For instance, if ⓡA is your dreaming that you win the lottery, (an R-accessible) w_1 is a fine world at which your dream comes true. The difference with the usual binary accessibility for modalities is in the broader set of accessible worlds: representation allows us to intend impossibilities.

The definitions of logical consequence and validity are standard. If S is a set of formulas:

---

via a number of reverse substitutions (which may be zero: a formula may already be its own matrix, if it has the proper structure). What happens at non-normal worlds is, in fact, the following: v assigns there to each matrix M of the relevant kind (a conditional or modal matrix at points in I, any matrix at points in E) pairs of subsets of D^n, that is, extensions and anti-extensions: if w is a non-normal world and M the relevant matrix, $v(M,w) = \langle v^+(M,w), v^-(M,w) \rangle$, with $v^+(M,w), v^-(M,w) \subseteq D^n$. Next, if $M[x_1,\ldots,x_n]$ is a matrix and $t_1, \ldots, t_n$ the susbstitutable terms, we have the following truth conditions for its substitution instances:

$w \Vdash_a^+ M[t_1,\ldots,t_n]$ iff $\langle v_a(t_1),\ldots,v_a(t_n) \rangle \in v^+(M,w)$
$w \Vdash_a^- M[t_1,\ldots,t_n]$ iff $\langle v_a(t_1),\ldots,v_a(t_n) \rangle \in v^-(M,w)$

See (Priest, 2005, pp. 26–9) for a proof that the matrix semantics works as expected. For the sake of brevity, I will keep talking of "treating complex formulas as atomic" at non-normal worlds; but it is this matrix procedure that is implied.

[6]Priest (2005) calls our extensionally impossible worlds *open worlds*, meaning that they are not closed under any non-trivial consequence relation; but they deserve to be called impossible if any world does.

$S \vDash A$ iff for every interpretation $\langle P, I, E, @, R, D, v \rangle$, and assignment a, if $@ \Vdash_a^+ B$ for all $B \in S$, then $@ \Vdash_a^+ A$.

As for logical validity:

$\vDash A$ iff $\emptyset \vDash A$, i.e., for every interpretation $\langle P, I, E, @, R, D, v \rangle$, and assignment a, $@ \Vdash_a^+ A$.

4 Two kinds of worlds

There are collateral, but philosophically interesting, reasons for flagging items in I among the non-normal worlds, that is, worlds less anarchic than those in E, where only the intensional logical vocabulary behaves in a deviant fashion. The distinction between intensionally and extensionally impossible worlds mirrors the presence of two positions in the current debate on the subject. The first may be labeled as the "Australasian stance". In the Australasian approach, worlds are constituents of interpretations of some relevant logic or other, which imposes to them some logical structure: they are closed under a relevant consequence relation, weaker than classical consequence relation (see e.g. Mares, 1997; Restall, 1997). Since this position draws especially on the conception of non-normal worlds as worlds where "logical laws may fail or be different", it is naturally allied to the idea that, at the (admissible) non-normal worlds, only intensional operators, such as a relevant conditional, behave in non-standard fashion. After all, it is the conduct of such operators that concerns the laws of logic. The truth conditions for conjunction, disjunction, or the quantifiers, should thus remain the same as in ordinary, possible worlds.[7]

The more radical view may be labeled the "American stance", since it reflects the opinion of some north-American impossible worlds theorists. The American stance focuses on the definition of non-normal worlds as "ways things could (absolutely) not be", and adopts what we may call an unrestricted comprehension principle for them. Roughly: for any way the world could not be, there is some impossible world which is like that. This can deliver particularly anarchic worlds, not closed under any non-trivial notion of logical consequence (see e.g. Vander Laan, 1997; Zalta, 1997).

---

[7] For similar considerations, see (e.g. Priest, 2001, ch. 9).

5 Relevant conditional

Having world quantifiers range on P ∪ I in the semantic clauses for $\to$ makes of it a relevant conditional, in the sense of fulfilling the aforementioned Variable Sharing Property. In particular, the arrangement above makes irrelevant entailments like $A \to (B \to B)$ fail—take a $w \in I$ where A is true but $B \to B$ is not. The failure is in the spirit of the "illogical" features of non-normal worlds: these are situations where laws of logic, like the law of sentential identity, may fail. EFQ as $(A \wedge \neg A) \to B$, and $A \to (B \vee \neg B)$, also fail (take a non-normal $w \in I$ where A is both true and false but B is untrue for the former, one where A is true but B is neither true nor false for the latter).

The conditional counts as a weak one by relevantist standards (it does not satisfy minimal contraposition, for instance). This may or may not be a problem, depending on what one expects from a conditional. A stronger setting can be obtained by adding to the interpretations for L a ternary relation on worlds and providing the semantics for a conditional in terms of it, as per the classical approach of (Routley & Meyer, 1973). This would complicate matters here, though. Our main concern is the representation operator ®, to which I now turn.

6 The (non-)logic of representation

The traditional debate in epistemic logic concerns the logical principles that should characterize the epistemic operators at issue, so as to mirror at best the corresponding intuitive notions. Some views are straightforward, for instance, knowledge being factive: if K_c is *cognitive agent c knows that*, it should sustain the entailment from $K_c A$ to A for any A.

Other inferences are more controversial. Must K_c allow the entailment from $K_c A$ to $K_c K_c A$? While this turns on issues concerning our intuitions about knowledge, it is not difficult to vindicate the inference, if we like it, by tampering with accessibility between worlds (in this case, just have it be transitive). But the failure of some basic logical inferences in epistemic and intentional contexts is more difficult to handle. This is the cluster of problems gathered under the well-known label of "logical omniscience". When modeled in standard possible world semantics, knowledge (or belief) turns out to be closed under entailment:

Non-Normal Worlds and Representation

(Cl) $A \to B, K_c A \vDash K_c B$

Also, all valid formulas turn out to be known (believed):

(Val) If $\vDash A$, then $\vDash K_c A$

And beliefs form a consistent set:

(Cons) $\vDash \neg(K_c A \wedge K_c \neg A)$

Taken together, these principles deliver an idealized notion of knowledge (belief), not mirroring the status of fallible and occasionally inconsistent cognitive agents.[8] Now Rantala's non-normal worlds were proposed to deal with these phenomena: despite being logically impossible, and not closed under any non-trivial consequence relation, they can be seen as viable epistemic alternatives by imperfect or inconsistent cognitive agents.

A similar story is to be told for ®. If we can conceive and represent impossibilities, the content of our representational state is the set of worlds that make our representation true, that is, where things are as they are conceived or represented to be; and this has to include non-normal worlds. Given the way things were set up above, non-normal worlds have no effect at the actual world @ on formulas not including ®. By allowing such worlds to be R-accessible in the evaluation of formulas including it, though, one can eliminate any unwelcome closure feature, thereby dispensing with (the formulations with ® in place of K_c of) (Cl), (Val), and (Cons).

As for (Cl), for instance: assume $\vDash A \to B$. Then at all worlds in $P \cup I$ where A holds, B holds. But there can be a non-normal world, w, at which A holds and B fails. If @ R w, then we can have that @ $\Vdash_a^+$ ®A, but it is not the case that @ $\Vdash_a^+$ ®B. Similarly for consistency: when the relevant R-accessible worlds are inconsistent worlds where both A and $\neg A$ are true, we can have @ $\Vdash_a^+$ ®$A \wedge$ ®$\neg A$.

7 Constraints

By accessing non-normal worlds of any kind on the one hand, and by not having constraints on its R-accessibility relation on the other, ®

---

[8]E.g. I know Peano's axioms as basic truths of arithmetic, and Peano's axioms entail (let us suppose) Goldbach's conjecture; but I do not know whether Goldbach's conjecture is true. With other intentional states such as belief or desire, also broad consistency is at stake.

has quite a poor logic—one may indeed wonder whether it is worth being called a *logic* at all. What is doing the interesting work here, though, is not the logic but the semantics. I am interested in the general form of the latter, and representability or conceivability had better be, generally speaking, quite anarchic.

In order to have Ⓡ express specific intentional operators under the generic umbrella of conceivability, say, *mentally representing a scenario* as opposed to *hallucinating*, we may nevertheless demand more structure. When one mentally represents a scenario, say, engaging in speculations on the next move of the financial markets, one's representation must have some more or less minimal coherence, that is, be closed under some, however weaker-than-classical, notion of logical consequence. This is proved by the fact that people meaningfully argue on how things are, and on what follows from what, in the relevant scenarios, that is, they accept or reject some things as holding in the situations at hand. Even when we represent to ourselves the impossible, we generally believe that we can draw inferences from what we explicitly represented.

One way to achieve this would be to place appropriate constraints on R-accessibility. We could then have Ⓡ model different species of representation depending on the constraints at issue. If there is something like *truthful representation* which is factive, we stipulate its R to be reflexive. Conversely, we may have *make-believe* representations such that the world w where the representing takes place is ruled out as a candidate for realizing them (as per the proviso to much fiction: "Any resemblance with real people or actual facts is merely accidental"). To have Ⓡ express something like "It is represented as holding *purely fictionally* that A", we stipulate R to be irreflexive.

Another way would be to make subdistinctions between non-normal worlds of various kinds. One may then allow only worlds that are closed under some form of entailment to be R-accessible, for instance, worlds in I. This gives us interesting results: representation then only accesses "typical" worlds of relevant logics, which are occasionally inconsistent or incomplete, and can also violate some logical laws, but are nevertheless adjunctive and prime (conjunction and disjunction behave standardly there). Then Ⓡ becomes closed under relevant entailment: if $\vDash A \to B$, and $@ \Vdash_a^+ ⓇA$, then $@ \Vdash_a^+ ⓇB$; thus, this kind of "relevant conceivability" brings a form of logical omniscience for *relevant* consequences of what is represented. However, inconsistent representation is still allowed, i.e., (Cons) fails, as well as (Val), i.e., not all logically valid formulae are represented.⁹

The need for further constraints is apparent when the representational act at issue is *fictional* representation, that is, the conceiving of situations described in fictional works, tales, stories, myths, etc. Sherlock Holmes is represented (at @), by Doyle and his readers, as a detective living in Baker Street, gifted with acute observational and logical skills, etc. Things are as they are represented at the worlds that make the relevant representational characterization true. But *which* are the relevant R-accessible worlds? That is: under which conditions does a world count as such that things are at it as they are represented? We want the relevant representations to be closed under some notion of logical consequence, so that if ®A, and B is a consequence of A, then ®B. In general, then, things represented in a certain way may well have *further* properties besides those they are explicitly represented as having. Some such properties will just follow on the basis of the entailments mandated by the logic for ®. For instance, from the fact that Tolkien represents Gandalf as a friend of Bilbo and Bilbo as a pipe-smoker, we can infer that Gandalf is represented as being friends with a pipe-smoker even though (let us suppose) Tolkien never says that explicitily.

On the other hand, what holds in a representation in many cases goes beyond both what is explicitly represented and what is entailed by logical implication. For while making inferences on what does or does not hold in a representation, we often import information from actuality, which we want to retain when assessing what goes on in a certain represented situation. What the relevant information is depends on our background knowledge of reality; but may also depend on our beliefs (even mistaken beliefs!). The import can rely on *ceteris paribus* and default clauses. Again, the case of fictional representation makes the point evident, and has been extensively studied, e.g., in (Lewis, 1978; Proudfoot, 2006). Doyle never explicitly represents (let us suppose) Holmes as living in Europe, or as having lungs. We are inclined to take these things as holding at all worlds that realize Doyle's characterization of Holmes, though, for we integrate the explicit representation with information imported from actuality. Now

---

⁹The closest antecedent to this in the literature, as far as I know, is Levesque's logic of explicit and implicit belief—see (Levesque, 1984).

Doyle certainly characterizes Holmes as a man living in London. At the actual world, London is in Europe and, if something is a normally endowed man, then it has lungs. Doyle says nothing against this, so, absent contrary indications from the author, the import is legitimate.

Intuitively, we should exclude from the R-accessible worlds that matter in evaluating what holds in the representation those worlds that, despite making true what is explicitly represented, add gratuitous changes with respect to actuality: we must exclude worlds that differ from @ more than required. Holmes is represented by Doyle as walking through London; we infer that Holmes is represented as walking through a European city. All worlds where Holmes walks through London but London is in Africa must be ruled out, for that would be a departure from actuality not mandated by what Doyle explicitly represents. London's being in Europe has to be held fixed across the worlds where things are as they are represented. This means that, to some extent, representations (of this kind) are about the real world as well. For what holds in a representation depends on what holds at the R-accessible worlds, where things are as represented. And which worlds these are depends also, to some extent, on how our reality is.

Even if this is worked out in a satisfactory way, it does not mean that we can expect precise answers to all the questions we may ask concerning a represented situation. Is Holmes, as characterized in Doyle's stories, right-handed or left-handed? Doyle does not say. And, intuitively, it is not the case that worlds where Holmes is left-handed in general differ gratuitously from @ more than worlds where he is right-handed, or vice versa. Representation typically under-represents.

Providing a detailed account of the workings of the representation operator, especially of how one is to select the worlds that are relevant to address what holds in a certain representation, is overall a difficult issue. Part of the difficulty is similar to the one of the standard treatment of counterfactuals *à la* Stalnaker-Lewis, where a counterfactual "If it were the case that A, then it would be the case that B" is true just in case the world(s) most similar to the actual world that make(s) the antecedent true, make(s) the consequent true as well. We need to invoke some notion of similarity between worlds, having to take into account worlds with minimal differences from actuality in certain respects. And this notion is notoriously slippery. The task becomes exceptionally tricky when we have to consider the intentions and beliefs of those who do the representing. Sometimes, for instance, an author of a work of fiction can make claims that, later on, turn out to be false in the story, or can make claims that are subtly ironic, etc.

What the appropriate constraints on R-accessibility are to be for the various species of representational activities is a difficult issue, and I am happy to leave it open here. Besides the similarities there is, in fact, a philosophical disanalogy between ⓡ and more traditional epistemic and intentional notions. That we are fallible and at times inconsistent as cognitive agents may be seen as a defect due to our finite and imperfect nature—when it's about knowing and, perhaps, believing. This is not so when it's about imagining and conceiving: in this case, logical fantasy is, generally speaking, a gift (or so I view it).

References

Brown, B. (1993). Old quantum theory: A paraconsistent approach. *Proceedings of the Philosophy of Science Association*, *2*, 397–441.

Levesque, H. (1984). A logic of implicit and explicit belief. *Proceedings of the National Conference on Artificial Intelligence (AAAI-84)*, 198–202.

Lewis, D. (1978). Truth in fiction. *American Philosophical Quarterly*, *15*, 37–46.

Mares, E. (1997). Who's afraid of impossible worlds? *Notre Dame Journal of Formal Logic*, *38*, 516–26.

Mares, E. (2004). *Relevant logic. A philosophical interpretation*. Cambridge: Cambridge University Press.

Priest, G. (2001). *An introduction to non-classical logic*. Cambridge: Cambridge University Press.

Priest, G. (2005). *Towards non-being. The logic and metaphysics of intentionality*. Oxford: Oxford University Press.

Proudfoot, D. (2006). Possible worlds semantics and fiction. *Journal of Philosophical Logic*, *35*, 9–40.

Rantala, V. (1982). Impossible worlds semantics and logical omniscience. *Acta Philosophica Fennica*, *35*, 106–15.

Restall, G. (1997). Ways things can't be. *Notre Dame Journal of Formal Logic*, *38*, 583–96.

Routley, R., & Meyer, R. (1973). The semantics of entailment. In

H. Leblanc (Ed.), *Truth, syntax and modality* (pp. 194–243). Amsterdam: North-Holland.

Vander Laan, D. (1997). The ontology of impossible worlds. *Notre Dame Journal of Formal Logic*, *38*, 597–620.

Zalta, E. (1997). A classically-based theory of impossible worlds. *Notre Dame Journal of Formal Logic*, *38*, 640–60.

Francesco Berto
Department of Philosophy and Northern Institute of Philosophy
University of Aberdeen
King's College, High Street, Aberdeen AB24 3UB, Scotland UK
e-mail: f.berto@abdn.ac.uk

The Evaluation Semantics—A Short Introduction

Frode Bjørdal

Abstract

We develop a philosophical alternative to the view that we need postulating a set W of possible worlds w_1, w_2, w_3, ... such that modal propositional logics are evaluated relative to valuations V, V', ... relative to the frame $\langle R, W \rangle$ where $R \subset W^2$. Here R is the accessibility relation on the first order entities in W. (The standard recursive clauses I take as known.) Instead let an evaluation E be a set of valuations V, V', V'', ..., and $R \subset E^2$ the accessibility relation. For given E and $R \subset E^2$, $\langle E, R \rangle$ is an *evaluation-frame*. Valuations in E obey Valuation-consistency (not both $V\alpha$ and $V{\sim}\alpha$), Valuation-completeness (either $V\alpha$ or $V{\sim}\alpha$), Valuation-disjunctivity ($V(\alpha \vee \beta)$ iff $V\alpha$ or $V\beta$) and Valuation-apodicticity ($V\Box\alpha$ iff for all $V'(VRV' \to V'\alpha)$). A semantic framework as adequate as possible worlds talk is induced. The generalization to quantified modal logics is straightforward though induces different correspondence conditions for the Barcan formula and its converse.

Keywords: alternative semantics, Barcan formula, evaluation semantics, modal logic, possible worlds, semantics

The purpose of the following is to come to more clarity concerning what entities we need commit to in order to provide adequate semantics for modal logics and related logics for e.g. counterfactuals. The theses defended have as consequence that we do not need possible worlds as first order objects, and certainly not pluralities so utterly implausibly suggested by David Lewis. As there is a plethora of points of view to make sense of what "possible worlds" **are**, e.g. modal fictionalism, it seems pertinent to point out certain simplifications that can be made in order to avoid some befuddlements which are prone to put philosophical discourse in disrepute. The view defended here is that we only need valuation-attributes, which may be regarded as states or properties of propositions (sentences) and, in the quantificational case, also of ordered pairs formed from elements of and sets from a given domain of discourse and linguistic items.

We first point out that we may presuppose a kindred semantics for classical propositional logic by letting valuations, or valuation-attributes, be states or properties of propositional variables of the formal language. For V a valuation and p a propositional variable, we have Vp or not Vp. For the recursive clauses we presuppose $V\sim p$ iff not Vp and $V(p \wedge q)$ iff Vp and Vq. It is straightforward to verify that α is a tautology iff $V\alpha$ for any valuation-attribute fulfilling the imposed constraints. A slight advantage with this approach over typical approaches is that we need not postulate functions that have truth or falsity, or more conventionally e.g. 0 or 1, as values when applied to formulas. The slight advantage lies in the fact that the denotations of "truth" and "falsity" remain obscure and the invocation of e.g. "0" and "1" seem arbitrary, whereas copulation as presupposed in the suggested framework here does not involve the postulation of such arbitrary or obscure entities.

We move on to modal logics and first consider *the propositional case*: Let Greek letters α, β, γ in the following stand for formulas in the object language of some modal propositional logic. Standardly, a formula α is seen as valid on a frame $\langle W, R \rangle$ iff $V(\alpha, w) = 1$ for all models $\langle W, R, V \rangle$ based on $\langle W, R \rangle$ and every $w \in W$. W is thought of as a set of *possible worlds*, and $R \subset W^2$ is the *accessibility relation* on W. Instead of $V(\alpha, w) = 1$, many prefer to write $wV\alpha$. This latter notation suggests that the "possible worlds" may be seen as functions from the value *assignment* V of a modal model to a compounded *valuation-property* wV. This hints that we instead of postulating a set of *possible worlds*, may restrict ourselves to what we call an *evaluation frame* $\langle \mathbf{E}, \mathbf{R} \rangle$, where the *evaluation* $\mathbf{E}$ is a set of *valuation-attributes* (often we just write *valuations*) V, V', ... and $\mathbf{R} \subset \mathbf{E}^2$ the accessibility relation on $\mathbf{E}$. We think of the evaluation frames as our *models*, and are interested in validity relative to various models, i.e. evaluation frames, with various restrictions on $\mathbf{R}$. In order to remind that we are making use of the evaluation semantics and not a standard possible worlds semantics we continually use the term "evaluation frame" instead of "model".

The Evaluation Semantics—A Short Introduction

Notice that what we henceforth think of as valuations in the evaluation semantics can neither be identified with possible worlds nor with valuation-assignments (often just called *valuations*) relative to frames. Instead, what we take as valuations in the evaluation semantics is rather some sort of hybrid, if you like.

For $V \in \mathbf{E}$ we write $V\alpha$ iff α holds according to valuation V. We require:

Valuation-consistency	If $V \in \mathbf{E}$ then not both $V\alpha$ and $V{\sim}\alpha$
Valuation-completeness	If $V \in \mathbf{E}$ then either $V\alpha$ or $V{\sim}\alpha$
Valuation-disjunctivity	If $V \in \mathbf{E}$ then $V(\alpha \vee \beta)$ iff $V\alpha$ or $V\beta$
Valuation-apodicticity	If $V \in \mathbf{E}$ then $V\Box\alpha$ iff $(\Pi V')(VRV' \rightarrow V'\alpha)$

In the last sentence we used another arrow for the material implication, and a capital Greek Π for the universal quantifier. This is used only occasionally here to underline that we are stating this in the metalanguage, and will be relaxed henceforth.

Exercise 1 Given only Valuation-consistency, Valuation-completeness, Valuation-disjunctivity and Valuation-apodicticity, show: a) Valuation-conjunctivity, i.e. for $v \in \mathbf{E}$, $V(\alpha \wedge \beta)$ iff $V\alpha$ and $V\beta$. b) Valuation-negativity, i.e. that for $V \in \mathbf{E}$, $V{\sim}\alpha$ iff **not** $V\alpha$. c) Valuation-implicativity, i.e. that for $V \in \mathbf{E}$, $V(\alpha \supset \beta)$ iff $V\alpha$ only if $V\beta$. d) Valuation-hypodicticity[1], i.e. that for $V \in \mathbf{E}$, $V\Diamond\alpha$ iff there is a V' such that VRV' and $V'\alpha$.

Result: A fully adequate semantics for modal propositional logics is induced, and soundness and completeness considerations carry over with few or none alterations. Restrictions on $\mathbf{R}$ induce different logics as in standard frameworks. This is left as an exercise here.

A note on modal correspondence is apt as an aside here. Typically correspondence results make appeals to a set theoretic apparatus (with lots of set theoretic comprehension) in order to obtain a correspondence between e.g. the modal schema $\Box\alpha \supset \alpha$ and the reflexivity of R. In our context this is not the way to go about. Instead we presuppose *linguistic comprehension principles* which put constraints on our evaluations. Let e.g. $\beth(V, V')$ be any first-order condition on the

---
[1] The term "hypodicticity" is a neologism; the term is philologically and etymologically reasonable and has its roots in the Greek term ὑποδειξις (hypodeixis).

valuations V and V' in R and $=$. We then presuppose that for all valuations V there is a β in the formal language so that for all V', $V'\beta$ iff $\beth(V,V')$. Let us with this consider the correspondence for e.g. reflexivity. It is elementary that $\Box \alpha \supset \alpha$ is valid in an evaluation frame ER $= \langle \mathbf{E}, \mathbf{R} \rangle$ when $\mathbf{R}$ is reflexive. Suppose $\Box \alpha \supset \alpha$ holds at all valuations $V \in \mathbf{E}$. We then have $V(\Box \alpha \supset \alpha)$ for all $V \in \mathbf{E}$. By exercise 1c), $V\Box\alpha$ only if $V\alpha$. Spelled out this gives that for all $V'(V\mathbf{R}V' \Rightarrow V'\alpha)$ only if $V\alpha$. By linguistic comprehension there is a β so that for all V', $V'\beta$ iff $V\mathbf{R}V'$. As α in the previous sentence is schematic, we substitute and obtain that for all $V'(V\mathbf{R}V' \Rightarrow V'\beta)$ only if $V\beta$. By linguistic comprehension this reduces to for all $V'(V\mathbf{R}V' \Rightarrow V\mathbf{R}V')$ only if $V\mathbf{R}V$. As the antecedent is a predicate logical tautology, we obtain $V\mathbf{R}V$. In consequence, $\Box\alpha \supset \alpha$ is valid in an evaluation frame ER $= \langle \mathbf{E}, \mathbf{R} \rangle$ iff $\mathbf{R}$ is reflexive. Other correspondence results are established similarly. The accommodation of second order conditions on valuations requires more linguistic comprehension to obtain suitable correspondence results. Linguistic comprehension for first order conditions will hold in all evaluations on account of the compactness of first order logic.

The quantificational case: We now let an *evaluation-model* be a triple $\langle \mathbf{D}, \mathbf{E}, \mathbf{R} \rangle$, where $\mathbf{D}$ is the domain of discourse, $\mathbf{E}$ the evaluation and $\mathbf{R}$ the accessibility-relation on $\mathbf{E}$. $\mathbf{E}$ is again a set of valuations V, V', ..., but now also on pairs as formed in the following. We assume $x, y, z, x', y', z', x'', \ldots$ etc. to be our variables and $F, G, H, \ldots$ etc. as predicates. We concentrate on the monadic situation as the generalization to the n-adic case is obvious. Valuation-consistency, Valuation-completeness, Valuation-disjunctivity and Valuation-apodicticity are still maintained as requirements. For $d \in \mathbf{D}$, we may have $V\langle x, d\rangle$ or not. If $V\langle x,d\rangle$, we write $V(x) = d$, as the relation is assumed to be functional. For a predicate F and D' a subset of $\mathbf{D}$, we may have $V\langle F, D'\rangle$ or not. Again we write $V(F) = D'$, as the relation on such pairs is again assumed to be functional. For relations, n-tuples are invoked as is standard. If x is a variable we write $V(x)V'$ to signify that V and V' at most differ in that there is one $d \in \mathbf{D}$ so that $V\langle x,d\rangle$ and not $V'\langle x,d\rangle$. We now impose plenist constraints on evaluations:

(1) Valuation-plenism: If $V \in \mathbf{E}$, x is a variable and $d \in \mathbf{D}$, $\exists V'(V' \in E\ \&\ V(x)V'\ \&\ V'(x) = d)$.

(2) Barcan-plenism: $\forall V', V''(V\mathbf{R}V' \& V'(x)V'' \to \exists V'''(V(x)V''' \& V'''\mathbf{R}V''))$.

(3) Converse-Barcan-plenism: $\forall V', V''(V(x)V' \& V'\mathbf{R}V'' \to \exists V'''(V\mathbf{R}V''' \& V'''(x)V''))$.

For formulas of monadic quantified modal logic we have

Valuation-atomicity: VFx iff $Vx \in VF$.
Valuation-generality: $V\forall x\alpha$ iff for all V' s.t. $V(x)V'$, $V'\alpha$.

Result: With the plenist constraints (1), (2) and (3), quantificational modal logics with the Barcan formula and its converse are straightforwardly accounted for. It is noteworthy that the Barcan-formula and its converse in the Evaluation Semantics are validated by imposing constraints which ensure that the evaluations considered are appropriate full, as stated by (2) and (3), and not by impositions concerning the domain. Completeness and soundness considerations carry over to our contexts with slight alterations. The constraints (1)–(3) may be adjusted to justify logics as in Kripke's approach with an underlying free logic. Notice that although only one domain is presupposed each valuation V may be assigned a sub-domain $\text{dom}(V) = \{d: d \in \mathbf{D} \ \&\ \text{there is a variable } x \text{ s.t. } V(x) = d\}$. If one wants, one may consider quantifier clauses relative to valuation specific sub-domains either in place of or as a supplement to Valuation-generality if one e.g. wants more than one style of quantifier. It is arguable that the evaluation semantics fits most naturally with the point of view that all objects exist eternally and necessarily, but more restrained quantifiers may supplement in as far as one seeks descriptions of more local concepts of reality.

The evaluation semantics is at least as flexible as standard possible worlds semantics, and as adequately, or perhaps better, accounts for modal discourse. Or at least so I argue. Details showing soundness and completeness are here left out as exercises and for more comprehensive forthcoming accounts.

Philosophical comments: It is obvious that there may be many set theoretic accounts of the semantics for modal logics, and I fully agree with Kripke in that there is no mathematical substitute for philosophy. However, attention to details is important. The evaluation semantics suggests that we indeed only need commit ourselves to *properties* (or, if one prefers to think more extensionally, *sets*) of sentences (propositions) and such pairs as have been invoked to semantically account for modal logics. (This is in deviation from the project of accounting for properties and propositions in terms of possible worlds, but that project has seemed misguided from its onset.) In the same way as there is no problem in accepting the existence of non-instantiated properties as *having won 30 Olympic gold medals*, there seems to be no ontological problem with accepting the existence of valuation-attributes which do not valuate propositions in accordance with how things really are. Notice well that our valuation-attributes should not be confused with *Ersatzworlds* as earlier suggested in the more philosophically bent literature. Residual perplexities concerning type are shared with standard approaches and not discussed here.

One could advantageously think of our valuation attributes as maximal consistent *states* which the world could be in. With such a terminology one is at liberty to think that the states invoked constitute second order analogues of so-called *possible worlds* in standard approaches. However, to my ears it sounds better to say that a statement/proposition/sentence is necessarily true if it holds in all possible states than to say that it holds in all (accessible) possible worlds. This is, one might say, perhaps because I am mundane. More seriously: I think it is true that our world could have been somewhat different from what it actually is. In the terminology which suggests itself from the Evaluation Semantics we may express this by saying that our world could have been in a different possible state. But I do not know how an adherent to possible world talk can express the idea that our world could have been in a different possible state, for to say that a different possible world could have obtained does not seem to assign a property to our world. Of course, here I have assumed that our world itself can be taken as a member of the domain **D**. It is unclear that adherents to standard possible worlds talk can assume naturally that there is one unique world to be presupposed semantically and that this world itself can be a member of the domain presupposed.

Frode Bjørdal
The Department of Philosophy, Classics and History of Art and Ideas
The University of Oslo, P.O.B. 1020 Blindern, Norway
e-mail: `frode.bjordal@ifikk.uio.no`
URL: http://www.hf.uio.no/ifikk/personer/vit/fbjordal/

On Dialetheic Entailment

Massimiliano Carrara   Enrico Martino
Vittorio Morato

Abstract

The entailment connective is introduced by Priest (2006b). It aims to capture, in a dialetheically acceptable way, the informal notion of logical consequence. This connective does not "fall foul" of Curry's Paradox by invalidating an inference rule called "Absorption" (or "Contraction") and the classical logical theorem called "Assertion". In this paper we show that the semantics of entailment, given by Priest in terms of possible worlds, is inadequate. In particular, we will argue that Priest's counterexamples to Absorption and Assertion use in the meta-language a dialetheically unacceptable principle. Furthermore, we show that the rejection of Assertion undermines Priest's claim that the entailment connective expresses the notion of logical consequence.

Keywords: dialetheism, entailment, possible world semantics, Curry's paradox

1 Introduction

In *In contradiction* (2006b), G. Priest has introduced a new connective aimed to capture, in a dialetheically acceptable way, the informal notion of *entailment* or *logical consequence*.

According to Priest, any dialetheically acceptable connective for entailment must, at least, have the following two features:

- it must validate MPP;

- it must not "fall foul" of Curry's Paradox.

An entailment connective must obey MPP because, according to Priest, it is the very meaning of the word "entailment" or "implication" to require it. It is then *analytically* true of any kind of genuine conditional that it should satisfy such a rule. "Any conditional worth its salt should satisfy the *modus ponens* principle", Priest writes. According to Priest, the classical material conditional → (or any conditional built on its basis, such as the strict implication, for example) cannot be used to capture the informal notion of entailment, at least in a dialetheic logical context. Priest holds that the material conditional is not *genuine*. In the logical framework of Priest (1979), where the material conditional was used, MPP was labelled a *quasi-valid rule*, a rule that is valid provided that all truth-values involved are classical (i.e., solely true or solely false).

An entailment connective must not "fall foul" of Curry's Paradox, because from Curry's Paradox follows *trivialism*, i.e., the thesis that every formula is true, and trivialism is to be avoided by the dialetheist.

Curry's paradox can be proved using self-reference devices, T-schemas, MPP, conditional proof and an inference rule called "absorption" or "contraction":

$$\text{ABS} \frac{\phi \to (\phi \to \psi)}{\phi \to \psi}$$

A dialetheic logic with entailment can avoid Curry's paradox, and thus trivialism, by invalidating ABS. This is in fact the strategy chosen by Priest.

It is to be noted, however, that there are other, quite standard, formulations of Curry's paradox that do not rely explicitly (i.e., at the level of the object language) on ABS. In some formulations, they rely on another related principle called "assertion" or even "pseudo *modus ponens*":

(Assertion) $\phi \land (\phi \to \psi) \to \psi$

As we will see, formulations of Curry's paradox with (Assertion) are blocked in Priest's approach in the same way in which formulations with ABS are blocked.

Formulations of Curry's paradox that do not rely on ABS at the level of the object language, however, typically make an appeal to such a rule at the level of the meta-language; in such cases, ABS is a *structural rule* governing the consequence relation.[1]

In Carrara, Gaio, and Martino (2010), criticisms to the entailment connective were done in the larger framework of criticizing the various attempts made by Priest to avoid trivialism generated by Curry's Paradox.

This article is a sequel of that work and will be squarely devoted to dialetheic entailment and its problems.

We are going to show that there is a tension between the logical rules used in the object language and those used in the meta-language.

2 The semantics of entailment

The characteristic feature of the entailment connective, that we will indicate from now on with the symbol $\Rightarrow$, is that of being a *modal* connective: its truth-conditions are in fact given in terms of a quantification over a set of possible worlds. Having a modal force is what distinguishes $\Rightarrow$ from the material conditional of classical logic. The modal force of $\Rightarrow$, however, is quite different from the force of other well-known modal conditionals, such as the strict conditional, or even the counterfactual conditional. Both conditionals, in fact, validate ABS and (Assertion).

An interpretation I for a language $\mathcal{L}$ with $\Rightarrow$ is given by a quadruple $\langle W, R, G, v \rangle$, where W is, as usual, an arbitrary set of objects ("possible worlds"), R is a dyadic relation between members of W ("the accessibility relation"), G is a designated member of W ("the actual world") and v a valuation function that assigns to each propositional atom and world w a non-empty subset of $\{0,1\}$, where 1 is the value "true", 0 is the value "false".

The semantic clauses that define the truth-conditions for a formula like $\phi \Rightarrow \psi$ are the following:

> $\phi \Rightarrow \psi$ is true in w if, and only if, for every world w' such that $R(w, w')$, if $1 \in v_{w'}(\phi)$, then $1 \in v_{w'}(\psi)$ and if $0 \in v_{w'}(\psi)$, then $0 \in v_{w'}(\phi)$

$\phi \Rightarrow \psi$ is then true in a world w if and only if, for every world w' accessible from w, if ϕ is true in w', so is ψ and if ψ is false in w', so

---
[1] Cfr. Beall and Murzi (in press).

is ϕ. From the clause above, it follows that a formula like $\phi \Rightarrow \psi$ is false at a world w if and only if there is at least one accessible world w' such that in w' ϕ is solely true and ψ is false or ϕ is true and ψ is solely false.

The definitions, respectively, of *semantic consequence* and *logical truth* are the following:

(SC) $\Gamma \models \alpha$ if and only if for all I, if, for every $\beta \in \Gamma$, $1 \in v_G(\beta)$, then $1 \in v_G(\alpha)$

(LC) $\models \alpha$ if and only if, for every I, $1 \in v_G(\alpha)$.

Note that the definitions of logical truth as truth in each actual world of every interpretation and of logical consequence as truth preservation in every actual world of every interpretation are in accordance with the standard Kripkean definitions of logical truth and of semantic consequence.

Counterexamples to ABS are obtained, as we will see, by means of interpretations with the following two features:

- G is *omniscient*: for every $w \in W$, $R(G, w)$

- R is non-reflexive: there is at least one $w \in W$ such that $\neg R(w, w)$

The omniscience of G means that G "sees" all other possible worlds; this implies that, for G, it holds that $R(G, G)$.

These kind of interpretations invalidate ABS. To prove that ABS is not a valid inference rule we must show that $\phi \Rightarrow \psi$ is not a semantic consequence of $\phi \Rightarrow (\phi \Rightarrow \psi)$. To do this, we have to show that there is at least one interpretation I such that $\phi \Rightarrow (\phi \Rightarrow \psi)$ is true at the actual world, while $\phi \Rightarrow \psi$ is false. Consider the following interpretation:

- $W = \{G, w\}$

- $R(G, w)$, $\neg R(w, w)$, $R(G, G)$, $R(w, G)$

- $v_G(\phi) = v_G(\psi) = v_w(\phi) = \{1\}$, $v_w(\psi) = \{0\}$

In such an interpretation, $v_G(\phi \Rightarrow (\phi \Rightarrow \psi)) = \{1\}$: since ϕ and ψ are true at G, the unique world accessible from w, $\phi \Rightarrow \psi$ is true in w. So in every world accessible from G, namely G and w, if ϕ is true, then $\phi \Rightarrow \psi$ is true.

In such an interpretation, however, $v_G(\phi \Rightarrow \psi) = \{0\}$, because there exists at least one world accessible from G (namely, w) where ϕ is true and ψ is false. Note that the use of $=$ instead of $\in$ for the evaluation function v signals that in such interpretation no dialetheia is involved. Counterexamples to ABS then have nothing to do with dialetheiae.

The very same interpretation can be used to show that (Assertion) is not valid. (Assertion) is invalid if, in some actual world, $\phi \wedge (\phi \Rightarrow \psi) \Rightarrow \psi$ is false in it. In the interpretation presented above, $v_G(\phi \wedge (\phi \Rightarrow \psi) \Rightarrow \psi) = \{0\}$, because there is at least one world accessible from G, namely w, where $v_w(\phi \wedge (\phi \Rightarrow \psi)) = \{1\}$ and $v_w(\psi) = \{0\}$.

The non-reflexivity of R and the omniscience of the actual worlds are essential for Priest's purposes. On the one hand, as we have just seen, if the R of the interpretations for $\Rightarrow$ were reflexive, ABS and (Assertion) would be valid. If, on the other hand, some Gs were not reflexive, then MPP would fail for $\Rightarrow$. Remind that failure of MPP for $\Rightarrow$ means that there is some actual world where ϕ and $\phi \Rightarrow \psi$ are true and ψ is false. Consider the following interpretation:

- $W = \{G, w\}$
- $\neg R(G, G)$, $R(G, w)$, $R(w, G)$, $R(w, w)$
- $v_G(\phi) = v_w(\phi) = v_w(\psi) = \{1\}$, $v_G(\psi) = \{0\}$

In this interpretation ψ is false in G, but ϕ and $\phi \Rightarrow \psi$ are true in G; in particular, $\phi \Rightarrow \psi$ is true in G because it is true in every world accessible to G, namely w.

The failure of (Assertion) shows that Priest's entailment fails to express logical consequence: while ψ is a semantic consequence of $\phi \wedge (\phi \Rightarrow \psi)$, this conjunction does not entail ψ.

The countermodels to ABS might be used to reveal other interesting characteristics of the language containing $\Rightarrow$ and, in particular, its interactions with the necessity operator $\Box$.

The fact that there is a non necessary formula where $\Rightarrow$ is the main connective, means that there is some formula that "logically follows" from another formula, but such that does not necessarily follows from it. This separation of logical necessity from logical consequence might be taken as undesirable: it is quite strange that α logically follows from β, but only contingently.

The contingency of entailment has consequences at the level of rules. As we already know, MPP is a valid rule in the logic of $\Rightarrow$. That MPP is a valid rule means that ψ is a semantic consequence of ϕ and $\phi \Rightarrow \psi$ and this means that in every actual world of every interpretation if $1 \in v_G(\phi)$ and $1 \in v_G(\phi \Rightarrow \psi)$, then $1 \in v_G(\psi)$. As our countermodel to ABS reveals, however, MPP "fails" in w, where $1 \in v_w(\phi)$ and $1 \in v_w(\phi \Rightarrow \psi)$ but $1 \notin v_w(\psi)$.

"Failure" of MPP in at least one non-actual world could be interpreted in two ways: on the one hand—and this would be perfectly compatible with the spirit of dialetheism—it could be taken as showing that non-actual possible worlds are deviant worlds; on the other hand—and we think that this is more problematic, even for the dialetheist—that *we*, from the standpoint of the actual world, are not able to reason about those worlds with our standard logical rules.

3 The philosophical justification of non-reflexivity and omniscience

What Priest aims to do in *In contradiction* is to give also a philosophical justification of omniscience and non-reflexivity. His views are revealed by this passage:

> Now, how do we know that all the "possible worlds" in an interpretation are conceivable by people living under those conditions of G? Simply because we are those people (by definition), and we conceive them. It is we who are theorizing, specifying what interpretations are, and we who can spell out any particular [assignment]. If we were to live under a different set of conditions, however, there would be no guarantee that we would be able to think all of this. Indeed, had we not evolved, we might have been maladapted to our environment, and might not even, therefore, have been able to conceive properly of the conditions under which we actually lived. G is omniscient, but there is no reason, therefore, why any other world should be omniscient or even reflexive. (Priest, 2006b, p. 87)

From this quoted passage we can extract the main motivations Priest uses to philosophically justify omniscience and non-reflexivity:

Omniscience of G: G is omniscient because the totality of possible worlds accessible from the actual world of an interpretation is the totality of the possible worlds conceivable by the inhabitants of G.

As we have seen, from the omniscience of G follows the reflexivity of G that Priest justifies in this way:

Reflexivity of G: G is accessible from itself because the inhabitants of G are "adapted" to their actual conditions and therefore they are able to conceive/represent them.

Finally, there is the non-reflexivity of some non-actual worlds from which it naturally follows their non-omniscience (if a non-actual world were omniscient, it would be accessible from itself).

Non-reflexivity and non-omniscience of non-actual worlds: We cannot grant inhabitants of other possible worlds the ability to conceive all possible worlds and even the ability to conceive their own situation.

There are various aspects of this philosophical picture that are problematic. We are going to mention just a few of them.

Against Omniscience, for example, it could be argued that it is generally assumed, in contemporary debates about the relations about conceivability and metaphysical possibility, that conceivability/possibility links fails in both directions;[2] it cannot then be assumed, or at least it cannot be assumed without argument, that a world is possible if and only if it is conceivable. We might not be able to conceive metaphysically or logically possible worlds that are accessible and thus relevant for the evaluation of sentences containing the entailment connective.

Furthermore, if omniscience of G is explained, as Priest does, via conceivability, there seems to be a clash between omniscience and non-reflexivity. Conceivability-based accounts of omniscience in fact seems to presuppose unrestricted reflexivity. Something is conceivable if it

---
[2]Cf. Gendler and Hawthorne (2002).

is conceived by us in an alternative (epistemic) situation. But to conceive something in an alternative situation we need to have access to this alternative situation. To know that we are conceiving something in another possible world (and therefore to know that something is conceivable by us in the actual world) we need to have access to the world in which we are conceiving. It could be denied that the reflexivity of a world has to be understood in terms of the capacity of its inhabitants to access this world; the quotation above reveals, however, that this is exactly what Priest had in mind.

Against non-reflexivity it could be argued that it does not fit well with the very nature of modal reasoning. The point of modal reasoning is that of reasoning in the actual situation about counterfactual situations. Without reflexivity, from knowing that ϕ is true in an accessible world w, we cannot even conclude that ϕ is possible in w (in non-reflexive frames, is false).

Another problem of non-reflexivity is revealed in the interpretations used to falsify ABS. In those interpretations, the evaluation of $\phi \Rightarrow \psi$ in w is done disregarding the truth-values of ϕ and ψ in w. For such an entailment to be true in w, what is relevant are just the truth-values of ϕ and ψ in G. This makes the evaluation of formulas containing $\Rightarrow$ quite different from the evaluation of formulas containing the other connectives.

But it is quite strange that the truth-values of ϕ and ψ are accessible when evaluating, for example, a conjunction in w, while they are not accessible when evaluating an entailment in w. What could be the interest of knowing the truth-value of $\phi \Rightarrow \psi$ in w if the truth-values of ϕ and ψ in w are just irrelevant to evaluate $\phi \Rightarrow \psi$ in w?

It seems therefore difficult to find an independent motivation for Omniscience and Non-reflexivity for the interpretation of $\Rightarrow$. These features seems to be specifically designed just to avoid ABS and therefore Curry's paradox.

In the next section we will show that, even from a logical point of view, the counterexamples to ABS are problematic.

4 Rejection and entailment

In *Doubt Truth to be a Liar* (2006a), Priest introduces the notion of *rejection*. The rejection of a certain proposition ϕ, according to

Priest, is a cognitive state consisting in the refusal of believing ϕ. To refuse to believe ϕ is having positive reasons to keep ϕ out of one's own belief box. The linguistic expression of rejection is *denial*. According to Priest, *in most contexts*, the assertion of a formula like $\phi \Rightarrow \bot$ constitutes the act of denial of ϕ.[3] In "normal conditions", writes Priest (2006a, p. 105), the rejection of ϕ could be expressed by the denial of ϕ, namely $\phi \Rightarrow \bot$.

What Priest intends by "most contexts" and "normal conditions" is explained by the following quotation:

> In most contexts, an assertion of [...] $\alpha \Rightarrow \bot$ would constitute an act of denial. Assuming that the person is normal, they will reject $\bot$, and so, by implication, α. The qualifier "in most contexts" is there because if one were ever to come across a trivialist who accepts $\bot$, this would not be the case. For such a person an assertion of $[\alpha \Rightarrow \bot]$ would not constitute a denial: nothing would. (Priest, 2006a, pp. 105–106)

From this it follows that the dialetheist is "normal", since he is not a trivialist; then he must accept that the rejection of ϕ could be expressed by $\phi \Rightarrow \bot$.

Now consider the Curry's sentence relative to $\bot$:

(Curry) $\phi \Leftrightarrow (\phi \Rightarrow \bot)$

If ϕ were true (possibly a dialetheia) then also $(\phi \Rightarrow \bot)$ would be true, due to the equivalence between the two; but then, by MPP, it would be possible to derive $\bot$. The dialetheist is thus forced, on pain of trivialism, to reject ϕ. If ϕ is rejected, however, also $(\phi \Rightarrow \bot)$ should be rejected. But then $(\phi \Rightarrow \bot)$ cannot be used to express the rejection of ϕ: in order to express the rejection of ϕ, $(\phi \Rightarrow \bot)$ must be true.

Furthermore, given that the rejection of ϕ implies the rejection of $(\phi \Rightarrow \bot)$, we must deny that $(\phi \Rightarrow \bot)$ is true, in spite of the falsity of the antecedent ϕ. In terms of the possible world semantics given for entailment, the falsity of $(\phi \Rightarrow \bot)$ implies the existence of a possible world w, accessible from the actual, where ϕ is true and $\bot$ is false.

---

[3] $\bot$ is a logical constant (*falsum*) such that it is a logical truth that $\bot \Rightarrow \alpha$, (for every α). $\bot$ is basically the symbol for an explosive sentence (i.e., a sentence implying all the others). $\bot$ must be solely false for the dialetheist, because if it were true, trivialism would follow.

Given that (Curry) is true in G, $(\phi \Rightarrow \bot)$ is true in w; but, for this entailment to be true in w, in every world w', accessible from w, it must happen that if ϕ is true, then $\bot$ is true. To define an interpretation where all these conditions obtain, we need to assign to ϕ the value 0 in every w' accessible to w; this in order to conclude that if ϕ is true then $\bot$ is true. But in order to justify this metalinguistic conclusion we need to accept, *in the metalanguage*, the derived rule of False Antecedent. It is only on the basis of the mere falsity of the antecedent "ϕ is true" that we can conclude that our metalinguistic statement is true.

The dialetheic justification for rejecting False Antecedent is that it would immediatly lead to trivialism. Assume that α is a dialetheia and consider an arbitrary formula β, or better $\bot$. In such a case α would be true (because a dialetheia is also true), but by False Antecedent also $\alpha \Rightarrow \bot$ would be true (because α is also false), but from this, by MPP, $\bot$ would be true and trivialism would follow.

In the case of (Curry), however, the dialetheist could defend the legitimacy of using False Antecedent for the metalinguistic proposition "if ϕ is true, then $\bot$ is true" because ϕ is not a dialetheia. But as we have seen above, Priest's solution to Curry's Paradox requires that, *in the object language*, $\phi \Rightarrow \bot$ be false in G even if ϕ is solely false. We have therefore a case where the sole falsity of an antecedent fails to guarantee the truth of the corresponding conditional. The falsity of $\phi \Rightarrow \bot$ is justified by Priest by the existence of a world accessible from the actual where ϕ is true and $\bot$ is false. But as we have seen above, in order to justify this, the rule of False Antecedent needs to be used in the metalanguage.

In the next section, we will argue that the use of False Antecedent in the metalanguage and the failure of False Antecedent in the object language conflicts with one of the main tenets of dialetheism, namely that the principles used in the metalanguage should be dialetheically acceptable.

5 Concluding remarks

Unlike the dialetheic solution to the Liar Paradox, Priest's solution to Curry's Paradox does not make use of dialetheias. The paradox is solved by postulating a genuine conditional, the entailment connective, that invalidates ABS, (Assertion) and solves Curry's Paradox.

The entailment connective is a modal one and its semantic is given in terms of possible worlds.

As we have already observed, failure of (Assertion) is in conflict with Priest's requirement that the entailment connective expresses the informal notion of logical consequence. This failure is essential to avoiding the paradox: trivialism is in fact a semantic consequence of (Curry) and (Assertion). For assume $\phi \Leftrightarrow (\phi \Rightarrow \bot)$ and $\phi \wedge (\phi \Rightarrow \bot) \Rightarrow \bot$ are true in G. Since $\phi \Rightarrow \bot$ is false in G, for the truth-conditions of entailment, there exists a possible, non trivial, world w where ϕ is true and $\bot$ is solely false. But in this world, the truth of ϕ implies the truth of $\phi \Rightarrow \bot$ (by (Curry)) and thus the truth of their conjunction $\phi \wedge (\phi \Rightarrow \bot)$, which is the antecedent of (Assertion). Therefore, $\bot$ must be true in w.

Counterexamples to ABS and Curry's Paradox, given in terms of such semantics, presuppose that a rule like False Antecedent be false for the entailment used in the object language, but true for the conditional used in the meta-language.

What kind of conditional is then used in the meta-language?

A dialetheically acceptable metalinguistic conditional, for which False Antecedent is correct, is the material conditional. But this, as we know, is not a genuine conditional (it invalidates MPP) and its use in the metalanguage would invalidate MPP also for $\Rightarrow$, on pain of trivialism, as the following interpretation shows:

- $W = \{G, w\}$

- no constraint on R

- $\{0, 1\} = v_G(\phi)$, $0 \in v_w(\phi)$

According to this interpretation, ϕ and $\phi \Rightarrow \bot$ are true in G, because in every world accessible to G, the meta-linguistic conditional "if ϕ is true, then $\bot$ is true" is true. Indeed, this proposition amounts to the disjunction "ϕ is false or $\bot$ is true", that is true in every world because ϕ is false; but then, by MPP, we obtain, in the object language, $\bot$.

On the other hand, if the metalinguistic conditional is a genuine conditional, it seems that, according to Priest, it should not satisfy False Antecedent. Thus it remains highly problematic for the dialetheist how to interpret the conditional used in the meta-language.

References

Beall, J. C., & Murzi, J. (in press). Two flavors of Curry paradox. *The Journal of Philosophy*.

Carrara, M., Gaio, S., & Martino, E. (2010). Can Priest's dialetheism avoid trivialism? In M. Peliš & V. Punčochář (Eds.), *The Logica yearbook 2010* (pp. 53–64). London: College Publications.

Gendler, T., & Hawthorne, J. (Eds.). (2002). *Conceivability and possibility*. Oxford: Oxford University Press.

Priest, G. (1979). The logic of paradox. *Journal of Philosophical Logic, 8*, 219–241.

Priest, G. (2006a). *Doubt truth to be a liar*. Oxford: Oxford University Press.

Priest, G. (2006b). *In contradiction*. Oxford: Oxford University Press.

Massimiliano Carrara, Enrico Martino, and Vittorio Morato
Department of Philosophy, University of Padua
P.zza Capitaniato, 3, 35139, Padua, Italy
e-mail:{massimiliano.carrara,enrico.martino,vittorio.morato}
    @unipd.it

The Asymmetry of Formal Logic

Colin Cheyne

Abstract

Formal logic is, in at least one important respect, asymmetric. The asymmetry is this: all substitution instances of valid argument forms are valid, but not all substitution instances of invalid argument forms are invalid. I discuss some implications of this asymmetry, in particular, an asymmetry that arises when formal logic is applied to the assessment of actual arguments, as well as discussing various misguided and unsuccessful attempts to avoid the implications.

Keywords: argument form, validity, invalidity, formal logic

Arguments and argument forms

By an *argument form* I shall mean a schema consisting of a string of symbols that are place-holders for either logical terms or non-logical (descriptive/content) terms: substituting terms of the appropriate kind for the symbols yields an argument. A *substitution instance* of an argument form is an argument that arises as a result of such a substitution. By a *valid* argument I shall mean an argument such that it is impossible for the premises to be true and the conclusion false. Thus, an *invalid* argument is an argument such that it is possible for the premises to be true and the conclusion false.

A valid argument form is an argument form of which, by definition, all substitution instances are valid. An invalid argument form is an argument form that is not valid, i.e. not all of its substitution instances are valid.

An example of a valid argument form is: $A \supset B, A \therefore B$.

A substitution instance of that form is:

> If Jack wins, then Betsy is happy,
> Jack wins,
> Therefore, Betsy is happy.

which is a valid argument.

An example of an invalid argument form is: $A \supset B, B \therefore A$.
A substitution instance of that form is:

> If Jack wins, then Betsy is happy,
> Betsy is happy,
> Therefore, Jack wins.

which is an invalid argument.

For the purposes of this paper, my examples will mostly be taken from propositional logic, although I conjecture that the claims I make with respect to asymmetry will apply to all other formal logical systems in common (and not so common) use, and irrespective of the notion of validity. However, I do not rule out the possibility of devising a logical system or notion of validity that avoids asymmetry. How useful such a devising might be would be another matter.

Note that my discussion is in terms of argument forms and their substitution instances. Other common ways of characterising the relationship between arguments and argument forms has it that the argument forms are interpretations or symbolisations or abstractions of the arguments. This approach (or cluster of approaches) gives rise to a number of problems and issues that I do not discuss (but see Peregrin & Svoboda in this volume).

The asymmetry

The asymmetry is easily illustrated. Consider the invalid argument form known as affirming the consequent:

$A \supset B, B \therefore A$

There are, of course, many invalid instances of this form, but consider these instances:

> If cats like cream then cats like cream,
> Cats like cream,
> Therefore, cats like cream.

The Asymmetry of Formal Logic

> If someone is happy, then Lucy is happy,
> Lucy is happy,
> Therefore, someone is happy.

These arguments are valid, so not all instances of this invalid form are invalid. To illustrate that this asymmetry is not just a feature of propositional logic, consider this invalid form in predicate logic:

> $(\forall x)(Px \supset Qx), Qa \therefore Pa$

and its valid substitution instance:

> All men are bachelors,
> Chris is a bachelor,
> Therefore, Chris is a man.

Another way of stating the asymmetry is to say that if an argument is a substitution instance of an invalid argument form, it does not follow that it is an invalid argument. This is easily proved. All valid arguments are instances of one of the following sequence of invalid argument forms:

> $\therefore C$,
> $A_1 \therefore C$,
> $A_1, A_2 \therefore C$,
> $A_1, A_2, A_3 \therefore C$,
> etc.

However, just as there are argument forms such that all of their substitution instances are valid, so also there are argument forms such that all of their substitution instances are invalid. These I shall call "super-invalid" argument forms. These super-invalid argument forms have premises that are formal tautologies and a conclusion that is a formal contradiction. For example:

> $P \vee {\sim}P \therefore Q \,\&\, {\sim}Q$

No matter what we substitute for P and Q, we will have an invalid argument, one with true premise and false conclusion.

The asymmetry of applied formal logic

Formal logical systems provide formal methods for proving the validity or invalidity of argument forms. Truth tables can provide such proofs in the case of propositional logic, but for other logical systems, proofs of validity are usually derivations and those of invalidity are counterexamples. Not all systems are complete in the sense that there is a proof for the validity of every valid argument form expressible in the system, let alone a proof of the invalidity of every invalid form. Furthermore, it is not always easy to find a proof even when there is one. However, it is an easy (mechanical) matter to establish that a given formal proof is sound and these formal proofs provide a high degree of rigour and certainty.

To sum up, we have formal methods for proving that:

(i) some argument forms are valid,

(ii) some argument forms are invalid,

(iii) some argument forms are super-invalid.

Once we have proved the validity of an argument form we can make an appropriate substitution to obtain a valid argument. A proof of the validity of that argument would consist of a formal proof plus a proof of the appropriateness of the substitution. The latter is not a formal matter. In fact, it could be quite problematic. One problem is that in order that the substitution be appropriate, terms substituted must have the semantics specified by the formal logical system. For propositional logic, the material conditional must be substituted for $\supset$, bivalent propositions for propositional symbols, and so on. Similarly, for predicate logic, vague predicates cannot be substituted for predicate symbols, etc. In other words, a substitution instance of an argument form may not, in spite of appearances, be an argument in natural language and may not be the argument which is of particular interest. However, any problems with the substitution step are the same whether the argument form is valid, invalid or super-invalid. I shall assume, for the purposes of this paper, that, at least in a wide range of cases, the substitution step is unproblematic. For convenience, I shall say that a proof of the validity of a form is a *formal method* for establishing the validity of an argument which is a substitution instance of that form.

The Asymmetry of Formal Logic

On the other hand, a proof of invalid form plus a substitution does not provide a formal method for establishing the invalidity of an argument, for the reason already discussed. Just because an argument is an instance of an invalid form, it does not follow that it is invalid. However, a formal proof of super-invalid form plus a substitution *does* provide a formal method for establishing the invalidity of an argument, since all instances of super-invalid forms are invalid.

Although super-invalid argument forms do not threaten the asymmetry thesis of formal logic, they do threaten an asymmetry thesis for the application of formal logic. We cannot simply say that formal logic provides formal methods for establishing the validity of arguments but not for establishing the invalidity of arguments. The invalidity of some, albeit very few, invalid arguments can be established formally, just as the validity of some, although not all, valid arguments can be established formally.

But there is something odd about the arguments whose invalidity can be established by a formal method. They are all substitution instances of super-invalid argument forms and, as such, have premises that are substitution instances of formal tautologies and a conclusion that is a substitution instance of a formal contradiction. Simple examples are:

> Snow is white or snow is not white,
> Therefore, the sky is blue and the sky is not blue.

> If grass is green then grass is green,
> Therefore, it is not the case that John is either rich or not rich.

These are not the sort of arguments we expect to arise in "real-life" situations. Indeed, I would go further and claim that such arguments almost never arise in real-life situations and are never likely to. (Notwithstanding, I should be delighted if anyone were to provide me with an actual example.) When it comes to applying formal logic to arguments that arise in real-life situations (what I shall call "everyday" arguments) we have the following asymmetry:

> Substitution instances of super-invalid argument forms are never (or scarcely ever) "everyday" arguments.

> Substitution instances of valid argument forms are often "everyday" arguments.

This, combined with the asymmetry thesis for formal logic, yields the important (albeit slightly messy) asymmetry thesis of applied formal logic.

> We can use a formal method (formal logic plus substitution) to establish the validity of many "everyday" arguments.

> We cannot use a formal method (formal logic plus substitution) to establish the invalidity of any "everyday" arguments.

Responses and evasions

Simple and obvious as this asymmetry is, it is often overlooked or neglected or even denied (directly or indirectly) by many logicians and philosophers. This is unfortunate, as the asymmetry has important implications for the teaching of logic, the epistemology of determining validity and invalidity, and the limitations of formal logic. Gerald Massey (1975, 1981) has been almost a lone voice in exploring some of these implications. Others who have made notable contributions on the issue are David Stove (1986, pp. 117–124) and Adrian Heathcote (1995). No doubt there are others. This is not to say that there aren't many other authors who explicitly or implicitly affirm the asymmetry, but do not see the need to discuss or defend it. Perhaps they see it as simple and obvious.

Attempts to deny or avoid or ameliorate the asymmetry are many and varied. I shall explore four representative examples here. They appeal, in turn, to the notions of (1) Logical Relatedness, (2) Logical Form, (3) Formal Invalidity, and (4) Tacit Premises.

Logical relatedness

The appeal to Logical Relatedness is an attempt to block valid arguments from being instances of an invalid form by placing a restriction on what counts as a legitimate substitution. The restriction is against the substitution of logically-related linguistic items for different symbols in an argument form. Items are logically related if any truth-value combinations are impossible. In the case of propositional logic, a pair

The Asymmetry of Formal Logic

of propositions are logically-related if it is impossible for them both to be true, both false, or one true and the other false. In the case of predicate logic, properties are logically-related if it is impossible for them both to instantiated, neither instantiated, or one instantiated and the other not.

For example, consider the invalid argument form:

$A \supset B, B \therefore A$

And the substitution instance:

> If someone is happy, then Betsy is happy,
> Betsy is happy,
> Therefore, someone is happy.

It is impossible for "Betsy is happy" to be true and "someone is happy" false, thus "Betsy is happy" and "someone is happy" are logically related and the substitution would be disallowed, blocking this valid argument as a substitution instance. On the other hand, the substitution:

> If Jack wins, then Betsy is happy,
> Betsy is happy,
> Therefore, Jack wins.

would be allowable because it is possible for "Betsy is happy" to be true and "Jack wins" false, and vice-versa, and for both to be true, and for both to be false, so they are not logically related.

This approach fails to overcome the asymmetry for at least two reasons. First, the judgement as to whether propositions (or properties, etc.) are logically related is a modal judgement equivalent to a judgement of invalidity. The asymmetry problem arises because we lack a formal method for making such judgements. An appeal to logical relatedness puts us no further ahead. Secondly, even if an appeal to logical relatedness were acceptable (*per impossibile*), asymmetry would not be eliminated because 'logically-related' substitutions into valid argument forms always yield valid arguments. So the restriction would only need to apply to substitution into invalid argument forms.

Logical form

The appeal to Logical Form also attempts to block valid arguments from being instances of an invalid form by placing a restriction on what counts as a legitimate substitution. The only allowable substitutions are into argument forms which exemplify *the* (true, unique?) logical form of the resulting argument. For example, the argument:

> If cats like cream then cats like cream,
> Cats like cream,
> Therefore, cats like cream.

can arise from substitution into both

$$A \supset B, B \therefore A$$

and

$$A \supset A, A \therefore A$$

but only the latter, it is claimed, exemplifies the argument's logical form. What constitutes an argument's true logical form is a complex issue that I shall not discuss here. Suffice to say that judgements of logical form are at least as demanding (probably more so) than judgements of validity and invalidity. Crucially, a necessary condition for correct logical form is that the form of a valid argument is a valid argument form and vice versa for an invalid argument. So the method proposed appears to be circular, since we need to determine whether or not an argument is valid in order to determine its logical form. As with the notion of logical relatedness, we are no further ahead. Furthermore, asymmetry is not eliminated because the restriction is not required for substitution into valid argument forms.

Formal invalidity

Rather than attempt to stipulate restrictions on what may count as allowable substitutions into logical forms, another approach is to claim that invalid arguments may be identified as those which do not arise from a substitution into *any* valid argument form. The first (obvious) problem with this approach is that just because we may not be able to specify a valid argument form for a given argument, it doesn't

The Asymmetry of Formal Logic

follow that there isn't one. So this approach cannot supply us with a formal method for proving validity. A weaker version suggests that we should regard any argument as invalid unless it can be shown to be a substitution instance of a valid argument form. (Guilty until proven innocent, as it were.)

A further problem, even for the weaker version, is that it assumes that all valid arguments have a valid argument form (i.e. are substitution instances of at least one valid argument form.) But this is by no means certain. Consider the valid argument:

> Chris is a bachelor,
> Therefore, Chris is a man.

which does not appear to have a valid argument form. One response to such arguments is an appeal to the notion of formal invalidity. Now, to say that an argument is formally valid usually means that it is valid in virtue of its form, in other words, it is the substitution instance of a valid argument form. Accordingly, the 'Chris is a bachelor' argument is not formally valid, but to suggest that it is therefore formally invalid is, at the very least, misleading. That would imply that it is invalid in virtue of its form, but that cannot be the case, because it is not invalid at all.

A further move (sometimes implicit, rather than explicit) is to re-define 'validity' as 'formal validity'. According to this new, 'improved' definition, any argument that is not (formally) valid becomes, by definition, invalid. Even this (desperate, but all too common) move will not rid of us asymmetry. It remains the case that we can prove by formal methods many "everyday" arguments' (formal) validity, but not the (formal) invalidity of any "everyday" arguments.

Tacit premises

A different approach to valid arguments of the 'Chris is a bachelor' variety is to claim that they *are* formally valid (i.e. that they are substitution instances of a valid argument form) because they have analytically true, but tacit, premises that render them so. The claim is that the argument:

> Chris is a bachelor,
> Therefore, Chris is a man.

tacitly assumes the analytic premise 'All bachelors are male' and is therefore an instance of the valid argument form:

$Pa, (\forall x)(Px \supset Qx) \therefore Qa$

On the other hand, the argument:

> Derek is bald,
> Therefore, Derek is intelligent.

is not formally valid, because there is no true analytic premise that renders it valid.

Even if we accept the dubious claim that the 'Chris is a bachelor' argument does have a tacit premise, we are faced with the problem of deciding whether or not certain propositions are analytic. There is no formal method to achieve this and in many cases it is highly controversial. For example, are the tacit premises in the following analytic or necessarily true?

> Eric is a cat,
> [All cats are mammals]
> Therefore, Eric is a mammal.
>
> This liquid is water,
> [Water is H_2O]
> Therefore, this liquid is H_2O.

Putnam (1962, p. 660) argues that it is possible that cats are robots planted on Earth by Martians, in which case they would not be mammals. Kripke (1980) argues that 'Water is H_2O' is necessarily true. These are both highly contested claims.

My sketch of typical responses to the asymmetry has been necessarily brief, but a clear pattern emerges. If we wish to appeal to argument form as part of a process to establish the invalidity of an 'everyday' argument, we will always need to do some extra, non-formal "stuff" that is not required to establish the validity of many (most) valid arguments. Even if we concede to some of suggestions made in response to the asymmetry, we are still left with a "modest" asymmetry thesis of applied formal logic:

> We can use formal logic to formally establish the validity of many "everyday" arguments.

The Asymmetry of Formal Logic

> We cannot use formal logic to formally establish the invalidity of any "everyday" arguments, without using "extra stuff".

I conclude with a look at some epistemological and pedagogical implications.

Epistemological implications

How do we come to know that an argument is invalid? Adrian Heathcote (1995, pp. 256–257) suggests that, because the asymmetry rules out a deductive method for establishing invalidity, we do it by a process of inductive reasoning. We search for a valid argument form that the argument instantiates. When we fail, we conclude that the best explanation for this failure is that the argument is invalid. This cannot be right. Such reasoning assumes that all valid arguments instantiate a valid argument form which is, at the least, dubious. Furthermore, it doesn't describe a method that is at all familiar. I, for one, cannot recall anyone explicitly arguing in this way. Sometimes, a person will demonstrate that the supposed form of an argument is invalid and conclude that the argument is invalid. Perhaps they implicitly assume that the argument has no other form that is valid, but perhaps they reason invalidly, in ignorance of the asymmetry.

In light of the asymmetry thesis, the only way to know or demonstrate that an argument is invalid is by direct appeal to the definition of invalidity. We must show that it is possible for the premises to be true and the conclusion false. Formal logic can assist with this, but it cannot provide the same degree of certainty. For (a very simple) example, the argument:

> If Jack wins, then Betsy is happy,
> Betsy is happy,
> Therefore, Jack wins.

can be shown to be a substitution instance of:

$A \supset B, B \therefore A$

The truth table for this argument form tells us that the premises would be true and the conclusion false if A were false and B true.

We can reach this point by formal method. But to conclude that the argument is invalid we need to know that it is possible for both 'Jack wins' to be false and 'Betsy is happy' to be true. This is a simple item of modal knowledge, knowable, perhaps, by modal intuition, or conceptual or linguistic understanding. However it is known, it is not by formal means. Furthermore, when it comes to such modal judgments in general, things are not always that simple. Consider, for example, the controversial possibility that Eric is a cat but not a mammal.

The means by which we can acquire knowledge of validity and invalidity differ. As a result, we can have knowledge of an argument's validity with something approaching the certainty with which we can have mathematical knowledge, but we cannot have knowledge of an argument's invalidity with the same degree of certainty.

Pedagogical implications

This difference in knowledge acquisition has pedagogical implications. What should we tell beginning students of logic? Not the following:

> So here's a strategy to use if you want to know whether an argument is invalid: First, ignore the content and isolate the logical form (the "skeleton") of the argument. [...] Second, see if you can invent an argument that has this logical form in which the premises are true and the conclusion is false. If you can find even one rotten apple of this type, you are finished. If there is even one argument with this property, then every argument of that form is invalid. (Sober, 2005, p. 12, his italics)

Similar statements can be found in other introductory texts. Many other texts are less explicit, but make it likely that students will wrongly infer a symmetry between valid and invalid forms. Still others offer an unsatisfactory response along the lines of those discussed above. Best, of course, are those that explicitly draw attention to the asymmetry. For example:

> If an argument has a Valid Form it is guaranteed to be valid, even if it has many invalid forms as well. But if an argument has an Invalid Form, there is *no guarantee* that

it is Invalid. So, while it is fairly straightforward to prove
that an argument is Valid, it is not easy to prove that it
is Invalid. (Girle, 2008, p. 135, his italics)

Having made the asymmetry clear, students should then be given examples of how to show that particular arguments are invalid. Typically, this will involve finding a counterexample to an invalid argument form and then showing or arguing that that counterexample is possible in the case of the argument in question, which instantiates the form. I illustrated this with a simple example above.

In conclusion: a question and speculative answer

My conclusion is that the asymmetry cannot be avoided. A question that arises is: Why do so many logicians and philosophers strive to ignore or avoid the asymmetry?

I believe that the quest for certainty has played a major role in creating an aversion to the asymmetry. We are able to devise formal logical systems within which we can determine the validity or invalidity of at least some argument forms with what appears to be mathematical certainty. By way of contrast, directly determining the validity or invalidity of arguments in natural language is much more uncertain. That relies on making modal judgments. We must decide whether it is possible or impossible for the premises to be true and the conclusion false.

Modal judgments are often difficult, controversial, and subjective. There is no mechanical procedure that can be applied or checked. Formal systems offer an attractive alternative, at least in the case of establishing validity. If we can show that a particular argument is a substitution instance of an argument form that has been proven valid, then we have established the argument's validity without an appeal to a particular (possibly uncertain and controversial) modal judgment.

The idea that we might use a similar method to establish invalidity is immediately attractive. Unfortunately, because of the asymmetry, showing that a particular argument is a substitution instance of an invalid argument form, does not suffice to show that the argument is invalid. Nor is there any further procedure that can conclusively demonstrate that a substitution instance of an invalid argument form is an invalid argument.

Formal logic promises certainty, but cannot deliver. We must resist its superficial allure. The uncertainty that surrounds the determination of invalidity is something that we must live with. In order to convince someone that an argument is invalid, we must convince them that it is possible for the premises to be true but the conclusion false. To merely point out that the argument instantiates an invalid argument form is not sufficient.

References

Girle, R. (2008). *Introduction to logic* (2nd ed.). North Shore, New Zealand: Pearson Education New Zealand.

Heathcote, A. (1995). Abductive inference and invalidity. *Theoria, 61*, 231–260.

Kripke, S. (1980). *Naming and necessity*. Cambridge, MA: Harvard University Press.

Massey, G. (1975). Are there any good arguments that bad arguments are bad? *Philosophy in Context, 4*, 61–77.

Massey, G. (1981). The fallacy behind fallacies. *Midwest Studies In Philosophy, 6*, 489–500.

Putnam, H. (1962). It ain't necessarily so. *The Journal of Philosophy, 59*, 658–671.

Sober, E. (2005). *Core questions in philosophy* (4th ed.). Upper Saddle River, NJ: Pearson Prentice Hall.

Stove, D. (1986). *The rationality of induction*. Oxford: Clarendon Press.

Colin Cheyne
Department of Philosophy, University of Otago
Box 56, Dunedin 9054, New Zealand
e-mail: `colin.cheyne@otago.ac.nz`

Reasoning by Analogy in Inductive Logic

Alexandra Hill*   Jeff Paris

Abstract

We propose a *Strong Analogy Principle* in the context of Unary Inductive Logic and characterize the probability functions which satisfy it. In particular, in the case of a language with just two predicates, the probability functions satisfying this principle correspond to solutions of Skyrms' 'Wheel of Fortune'.

Keywords: uncertain reasoning, inductive logic, analogy principle, analogical reasoning, wheel of fortune

Introduction

The key tenet of inductive reasoning is that the future is likely to resemble the past. To put this another way, our expectation, or rational probability, of a given event occurring will be positively affected by having seen similar events in the past. Clearly then, an analysis of what it is for events to be similar is important for the development of Inductive Logic. Of course, two identical events are certainly similar, and this idea gives rise to the most basic precept of Inductive Logic, the Principle of Instantial Relevance:

The Principle of Instantial Relevance (PIR)
If w is a rational probability function on a language with constants $\{a_1, a_2, a_3, \ldots\}$ then

$$w(\phi(a_{j+1}) \mid \phi(a_j) \wedge \theta) \geq w(\phi(a_{j+1}) \mid \theta)$$

for any formula $\phi(x)$ and set of observations θ not involving a_j, a_{j+1}.

---

*Supported by a UK Engineering and Physical Sciences Research Council (EPSRC) Research Studentship.

(PIR) says, roughly, that your expectation of seeing a particular event should be positively affected by having seen an identical one in the past. (PIR) can be thought of as a minimal requirement for a probability function to satisfy before it can be said to support inductive reasoning, but it is still relatively crude. In the real world, most of our inductive reasoning concerns events which are similar but non-identical, and yet Inductive Logic has not been able to give a satisfactory account of this kind of situation. The aim of this paper is to describe some attempts at modeling arguments by analogy and suggest our own Strong Analogy Principle based on a comparative notion of similarity. Before we do so, let us consider a few motivating examples.

Consider the person who knows that the vast majority of venomous snake have slanted, cat-like eyes whereas the vast majority of non-venomous snakes have round eyes and pupils. If she comes across a snake unlike any she has seen before, but with slanted eyes, it would be supremely rational to stay well away from it! Even though she has never seen an identical creature before, it shares a sufficient number of properties with those in her experience for her to infer the likelihood of it sharing the further property of being venomous. Another example: suppose a person has only ever seen green peppers and red peppers, and dislikes the taste of the former but likes the taste of the latter. Presented with an orange pepper and a yellow pepper, we expect them to be more inclined to try the orange one, it being more similar to their favourite red.

These examples demonstrate two ways in which two objects, A and B, can be similar to one another:

1. A and B are known to share a number of *identical properties* (but may not share all their properties).

2. A and B are known to share a number of *similar* properties (perhaps as well as some other ones).

The first example explains similarity in terms of the more basic notions of identity and non-identity, whereas the second requires some further explanation of what it is for two properties to be similar. For present purposes we will focus on examples of the former kind, although we believe that it is possible to explicate a notion of distance between properties and from there give a precise explanation of examples of the second kind. Needless to say, this is more complicated and so we restrict our attention here to examples of the simpler kind.

Context and notation

To formalize our ideas we work in the conventional (unary) Inductive Logic context. That is, we have a predicate language L_q without equality or functions and having finitely many predicates, $P_1, \ldots, P_q$, and constants, $a_1, a_2, a_3, \ldots$, the intention being that these enumerate the universe.[1] Let SL_q denote the set of sentences of L_q and $QFSL_q$ the quantifier free sentences of L_q and similarly $FL_q/QFFL_q$ for formulae. We define the *atoms* of L_q, which we denote by $\alpha_i(x)$ $i = 1, 2, \ldots, 2^q$, as the 2^q formulae of the form

$$\pm P_1(x) \wedge \pm P_2(x) \wedge \ldots \wedge \pm P_q(x)$$

We further define *state descriptions* to be sentences of the form

$$\alpha_{i_1}(a_1) \wedge \alpha_{i_2}(a_2) \wedge \ldots \wedge \alpha_{i_n}(a_n)$$

which we can abbreviate as $\bigwedge_{i=1}^{2^q} \alpha_i^{n_i}$ where n_i is the number of times α_i appears as a conjunct.

In this context, we define a Probability Function on L_q as follows.

Definition 1 A Probability Function on L_q is a map $w \colon SL_q \to [0, 1]$ such that for all $\theta, \phi, \exists \psi(x) \in SL_q$:

(P1) If $\models \theta$ then $w(\theta) = 1$.

(P2) If $\models \neg(\theta \wedge \phi)$ then $w(\theta \vee \phi) = w(\theta) + w(\phi)$.

(P3) $w(\exists x \psi(x)) = \lim_{m \to \infty} w(\bigvee_{i=1}^{m} \psi(a_i))$.

Our interest in Inductive Logic is to pick out probability functions on L_q which are arguably *logical* or *rational* in the sense that they could be the choice of a rational agent. Or to put it another way to discard

---

[1] Some authors, following Carnap (see for example Carnap, 1971), allow families of mutually exclusive and jointly exhaustive predicates $\{P_1^1, \ldots, P_n^i\}$ in the place of the two membered families $\{P_i, \neg P_i\}$. This may be advantageous for explicating the more complicated kind of similarity mentioned in the introduction, but since out intention here is that all predicates be considered equally similar or dissimilar to one another, it is not necessary for us to impose this further structure.

probability functions which could be judged in some sense to be 'irrational'.[2] The usual method of thinning down towards such rational choices is to impose 'rationality principles' which these probability functions should arguably satisfy. Of course there can be considerable disagreement about which principles are rational (to the extent of different candidates being mutually inconsistent, see for example Nix & Paris, 2006) but one such widely accepted principle is that the inherent symmetry between the constants should be respected by any rational probability function w on L_q. Precisely w should satisfy:

The Constant Exchangeability Principle (Ex)
For $\theta, \theta' \in QFSL_q$, if θ' is obtained from θ by replacing the constant symbols[3] $a_{i_1}, a_{i_2}, \ldots, a_{i_m}$ in θ by $a_{k_1}, a_{k_2}, \ldots, a_{k_m}$ respectively, then $w(\theta) = w(\theta')$.

In what follows we take Ex as a standing assumption.

Extending the idea of symmetry between symbols of the language, we might also feel it rational to require that the predicates are exchangeable.

The Predicate Exchangeability Principle (Px)
For $\theta, \theta' \in QFSL_q$, if θ' is obtained from θ by replacing the predicate symbols $P_{j_1}, P_{j_2}, \ldots, P_{j_m}$ in θ by $P_{s_1}, P_{s_2}, \ldots, P_{s_m}$, respectively, then $w(\theta) = w(\theta')$.

A further principle suggested by the idea of symmetry is that of strong negation:

The Strong Negation Principle (SN)
For $\theta, \theta' \in QFSL_q$, if θ' is obtained from θ by replacing each occurrence of $\pm P_i$ by $\mp P_i$ for some predicate P_i, then $w(\theta) = w(\theta')$.

We shall assume throughout that the probability functions we consider satisfy Px, SN

The Principle of Regularity (REG)
For consistent $\phi \in QFSL$, $w(\phi) \neq 0$.

---

[2] It is interesting that 'irrationality' seems much easier to spot than 'rationality'.

[3] In lists like this we will always assume that the entries are distinct unless stated otherwise.

As will be clear shortly the assumption of regularity is implicit in our forthcoming Strong Analogy Principle so it only clarifies the situation to make it explicit from the start. Restricting to such probability functions will be convenient because it ensures that conditioning on consistent $\phi \in QFSL_q$ is well defined. And in any case the rationality of Regularity is hardly a point of widespread contention asserting as it does that if $\phi \in QFSL_q$ is not impossible then it should be assigned some non-zero probability.

Before we detail our Strong Analogy Principle (SAP) we shall briefly review some other attempts at capturing 'analogical influence'.

Central to the search for logical probability functions has been Carnap's proposed *Continuum of Inductive Methods*, the c_λ^q probability functions on L_q (see for example Carnap, 1971, 1980). For $0 < \lambda \leq \infty$, c_λ^q is characterized by the special values

$$c_\lambda^q \left(\alpha_j(a_n+1) \mid \bigwedge_{i=1}^n \alpha_{h_i}(a_i) \right) = \frac{n_j + \lambda 2^{-q}}{n + \lambda}$$

where n_j is the number of occurrences of α_j amongst the α_{h_i}, whilst for $\lambda = 0$ c_0^q is characterized by

$$c_0^q \left(\bigwedge_{i=1}^n \alpha_{h_i}(a_i) \right) = \begin{cases} 2^{-q} & \text{if all the } h_i \text{ are equal,} \\ 0 & \text{otherwise.} \end{cases}$$

We will refer to functions of this kind as c_λ-functions. This approach can also be generalized to give what we shall refer to as $c_{\lambda\gamma}$-functions, where γ is any prior distribution $\langle \gamma_1, \ldots, \gamma_{2^q} \rangle \in \mathbb{D}_q$, and

$$c_{\lambda\gamma}^q \left(\alpha_j(a_n+1) \mid \bigwedge_{i=1}^n \alpha_{h_i}(a_i) \right) = \frac{n_i + \lambda \gamma_j}{n + \lambda}$$

These functions satisfy (PIR) but do not support any analogy based inferences. Carnap himself viewed this as a problem and (Carnap & Stegmüller, 1959) proposes an Axiom of Analogy—very similar to our forthcoming proposed analogy principle SAP but with restrictions on the evidence that can be conditioned on.[4] The suggestion there is that

---

[4]In fact he suggests a series of Axioms of increasing strength, the strongest of which is equivalent to SAP, but does not go so far as to classify the functions satisfying any of these.

it is satisfied by a mixture of c_λ-functions weighted by a parameter η, which can be viewed as measuring the strength of the analogy effects. While this approach does generate some of the desired effects it also violates his Axiom of Analogy (and SAP) in some cases. A slightly different Principle of Analogy appears in (Carnap, 1980) but the class of probability functions satisfying it is not derived there.

Much subsequent work in Inductive Logic proceeds by attempting to modify or combine c_λ-functions. So, for example, Skyrms (1993) combines Carnapian functions to give a probability function which exhibits some limited properties of analogical reasoning. He uses the illustrative example of a Wheel of Fortune with four outcomes—North, East, West and South, outcomes which we will model, respectively, by the four atoms in L_2, $\alpha_1(x) = P_1(x) \wedge P_2(x)$, $\alpha_2(x) = P_1(x) \wedge \neg P_2(x)$, $\alpha_3(x) = \neg P_1(x) \wedge P_2(x)$, $\alpha_4(x) = \neg P_1(x) \wedge \neg P_2(x)$. In our notation, Skyrms proposes the probability function

$$w = 4^{-1} \left(c^2_{\lambda \gamma_n} + c^2_{\lambda \gamma_e} + c^2_{\lambda \gamma_w} + c^2_{\lambda \gamma_s} \right)$$

where

$$\gamma_n = \left\langle \frac{1}{2}, \frac{1}{5}, \frac{1}{5}, \frac{1}{10} \right\rangle, \quad \gamma_e = \left\langle \frac{1}{5}, \frac{1}{2}, \frac{1}{10}, \frac{1}{5} \right\rangle,$$

$$\gamma_w = \left\langle \frac{1}{5}, \frac{1}{10}, \frac{1}{2}, \frac{1}{5} \right\rangle, \quad \gamma_s = \left\langle \frac{1}{10}, \frac{1}{5}, \frac{1}{5}, \frac{1}{2} \right\rangle.$$

It can be shown that $w(n \mid n) = 22/55$, $w(n \mid e) = w(n \mid w) = 12/55$ and $w(n \mid s) = 9/55$, where $w(n \mid n)$ is the probability of a north on the second spin of the wheel given that the first spin had already landed north, etc., so in this case at least the required inequalities hold.[5]

However, the analogy effects of Skyrms' function are very limited. In fact they break down very quickly when the probabilities are conditioned on more than one prior observation. For example, Festa (1996) shows that for $\theta = ssssswwwww$, $w(n \mid e \wedge \theta) < w(n \mid s \wedge \theta)$, contradicting the expectations of our rational agent (and our forthcoming principle SAP).

---

[5] Skyrms actually considers $w(n \mid n)$, $w(e \mid n)$, $w(w \mid n)$ and $w(s \mid n)$ but since these are respectively equal to $w(n \mid n)$, $w(n \mid e)$, $w(n \mid w)$, $w(n \mid e)$ this amounts to the same thing.

Festa, having demonstrated the limitations of Skyrms' proposed function, does generalize Skyrms' method in an attempt to improve the scope of the analogy effects displayed there. He does so successfully for a hypothetical situation in which we know that for some ordered triple $\langle \alpha_i, \alpha_j, \alpha_h \rangle$, α_j is more similar to α_i than α_h is, while α_i and α_j are equally similar to α_h. In other words, the distance between the three predicates can be pictured thus

However, his method is not able to deal with a situation in which there is more than one such triple.

Niiniluoto (1981) and Kuipers (1984) proceed by weighting $c_{\lambda\gamma}^2$ functions, including a further parameter in the same spirit as Carnap's η. We mention these only briefly as while this approach ensures some of the desired analogical properties, the resulting functions violate Ex, which we regard as a fatal failing.

Maher (2001) and Romeijn (2006) also use Carnapian methods to capture analogy effects in a language with four atoms, but unlike the above mentioned authors their models make use of underlying predicate families. Maher's method is very similar to Carnap's first suggestion in (Carnap & Stegmüller, 1959), the only difference being that in place of Carnap's parameter η he uses the probability $w(I)$ where I is the proposition that the two predicates are stochastically independent. The resulting function only satisfies AP in some restricted cases, but Maher does not suggest an alternative principle which they do satisfy.

Romeijn also defines functions on predicates rather than directly on atoms and the resulting function satisfies a principle equivalent to a generalization of the principle SAP which we shall propose. However, as Romeijn notes, one major limitation of this model is the asymmetry with which it treats predicates. In particular his function does not satisfy Px. As suggested previously, the failure of Px might be acceptable if there were a similarity relation defined directly on predicates which also affected the analogies between atoms. However Romeijn's language is identical to our L_2 and so there seems to be no justification for an asymmetry between the predicates. While the failure of Px may not be a fatal flaw, it seems to do so here in a particularly arbitrary way. It is not clear from Romeijn's work precisely how he conceives of similarity or how he wishes to treat the primitive predicates.

All of these investigations into analogy take what we consider to be a rather inefficient approach. That is, rather than investigating any principle directly, they investigate candidate probability functions. This means that even if it can be demonstrated what kind of analogy influence a function allows for, we are none the wiser as to the full class of functions showing such a property. In fact, many of the above examples do not go so far as to show precisely what kind of analogy principle is satisfied anyway, rather they give a handful of examples. It is our view that it is most illuminating to investigate the consequences of of a precisely stated analogical principle directly and see what probability functions it admits.

The Strong Analogy Principle

The idea behind our proposed Strong Analogy Principle is that if $\theta(x), \chi(x), \psi(x) \in FL_q$ and $\theta(x)$ is 'nearer' to $\chi(x)$ than $\psi(x)$ is then, whatever our previously acquired knowledge ϕ might be, learning $\theta(a_{n+1})$ should increase our probability in $\chi(a_{n+2})$ more than learning $\psi(a_{n+1})$ would have done, no matter what we had previously learnt. Before this idea can be transformed into a precise principle however there are various issues here that need to be clarified.

Firstly our previously acquired knowledge should obviously not directly involve a_{n+1} or a_{n+2}. For this reason we assume that it is of the form $\phi(a_1, a_2, \ldots, a_n) \in QFSL_q$.

Secondly what do we mean by 'near'? As discussed in the introduction, we are thinking of similarity as deriving from the sharing of identical properties, which in the case of nearness relation between the atoms of the language L_q leads naturally to measuring it by Hamming distance:

Definition 2 The Hamming Distance between two atoms, α_i and α_j, is the number of $\pm P_i$ in which they differ. We denote the Hamming

Distance by $|\alpha_i - \alpha_j|$.

In the case of L_2, for the four atoms $\alpha_1, \alpha_2, \alpha_3, \alpha_4$ described earlier, the Hamming Distances between the atoms can be pictured thus:

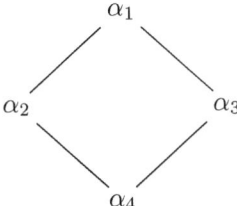

Note the appropriateness of this framework to model the situation that Skyrms is interested in, with the four atoms associated with the four outcomes of the the Wheel of Fortune. These considerations now suggest a very natural analogy principle:

The Strong Analogy Principle (SAP)
For atoms $\alpha_i, \alpha_j, \alpha_k$, if $|\alpha_i - \alpha_j| < |\alpha_i - \alpha_k|$ then

$$w(\alpha_i(a_{n+2}) \mid \alpha_j(a_{n+1}) \wedge \phi(a_1, a_2, \ldots, a_n)) > \\ w(\alpha_i(a_{n+2}) \mid \alpha_k(a_{n+1}) \wedge \phi(a_1, a_2, \ldots, a_n)) \quad (1)$$

for any consistent $\phi(a_1, a_2, \ldots, a_n) \in QFSL_q$.[6]

To avoid an excess of notation we may in future applications abbreviate expressions such as (1) to

$$w(\alpha_i \mid \alpha_j \wedge \phi) > w(\alpha_i \mid \alpha_k \wedge \phi)$$

the distinct instantiating constants being left implicit (we are entitled to do this since by Ex it does not matter which constants appear as long as they are all distinct).

We call the above principle the Strong Analogy Principle in virtue of the strict inequality in (1), and to distinguish it from what we have in (Hill & Paris, n.d.) simply called the Analogy Principle (AP) where the inequality is taken to be greater or equal. The only extra functions

---

[6]Recall that we are assuming throughout that w satisfies Regularity so these conditional probabilities are well defined.

permitted by AP are trivial solutions, so it is of more interest to look at SAP here. Full details of this may be found in (Hill & Paris, n.d.).

In what follows we shall give a complete characterization of those probability functions satisfying SAP, in the case of a language with just one or two or more predicates, in the presence of Ex, Px, SN[7] and REG. Our first proposition concerns the language L_1 where there is only a single predicate. Recall that for this language Carnap's probability functions c_∞^1 and c_0^1 are defined as follows:

$$c_\infty^1 \left(\bigwedge_{i=1}^n \alpha_{h_i}(a_i) \right) = 2^{-n}$$

$$c_0^1 \left(\bigwedge_{i=1}^n \alpha_{h_i}(a_i) \right) = \begin{cases} 2^{-1} & \text{if all the } h_i \text{ are equal,} \\ 0 & \text{otherwise.} \end{cases}$$

Proposition 1 *Suppose that the probability function w on L_1 satisfies Ex, Px, SN and REG. Then either $w = \lambda c_\infty^1 + (1-\lambda)c_0^1$ for some $0 < \lambda \leq 1$ or w satisfies SAP.*[8]

We now turn to L_2.

For $\vec{b} \in \mathbb{D}_q$ define $w_{\vec{b}}$ to be the probability function on L_q given by

$$w_{\vec{b}} \left(\bigwedge_{i=1}^m \alpha_{h_i}(a_{r_i}) \right) = \prod_{i=1}^m b_{h_i} = \prod_{j=1}^{2^q} b_j^{n_j}$$

where n_j is the number of times that α_j appears amongst the α_{h_i}.

For $\langle a, b, c, d \rangle \in \mathbb{D}_2$ let $y_{\langle a,b,c,d \rangle}$ be the probability function given by

$$y_{\langle a,b,c,d \rangle} = 8^{-1}(w_{\langle a,b,c,d \rangle} + w_{\langle a,c,b,d \rangle} + w_{\langle b,a,d,c \rangle} + w_{\langle c,a,d,b \rangle} \\ + w_{\langle b,d,a,c \rangle} + w_{\langle c,d,a,b \rangle} + w_{\langle d,b,c,a \rangle} + w_{\langle d,c,b,a \rangle}).$$

Note that the eight summands correspond to the permutations of α_1, α_2, α_3, α_4 that preserve Hamming Distance, where the atoms are

---

[7] Notice that it would seem 'irrational' not to assume these two symmetry principles Px and SN since the permutations of atoms that they correspond to preserve Hamming Distance. Indeed, see (Hill & Paris, n.d.), these are exactly the permutations of atoms which preserve Hamming Distance.

[8] The proof of this and the proofs of all further results we mention will appear in (Hill & Paris, n.d.).

associated with a, b, c and d respectively. We are now in a position to classify the probability functions satisfying AP on L_2 with the following theorem.

Theorem 2 *Let w be a probability function on L_2 satisfying Ex, Px, SN, REG and SAP. Then one of the following hold:*

(1) $w = y_{\langle a,b,b,c \rangle}$ for some $\langle a,b,b,c \rangle \in \mathbb{D}_2$ with $a > b > c > 0$, $ac = b^2$.

(2) $w = \lambda y_{\langle a,a,b,b \rangle} + (1-\lambda)c_\infty^2$ for some $\langle a,a,b,b \rangle \in \mathbb{D}_2$ with $a > b > 0$ and $0 < \lambda \leq 1$.

Having seen that we have quite a narrow class of functions satisfying SAP on L_2 our next result shows that the situation is even more bleak for larger languages.

Theorem 3 *If w is any probability function on L_k for $k \geq 3$ satisfying Ex, Px, SN and REG then w does not satisfy SAP.*

Conclusions

It is disappointing that SAP, which has such intuitive rationality, is too strong to generate a class of probability functions on languages with more than two predicates. As mentioned, even the weaker AP only generates trivial solutions, see (Hill & Paris, n.d.). Since reasoning by analogy seems to be such a fundamental feature of inductive arguments, the failure of SAP suggests that we should next consider ways of modifying our principle of analogy. One idea is to place restrictions on the past evidence to be conditioned on. Above we allowed any sentence to appear as past evidence (and this is used at crucial points in the proof of Theorem 2), but if we restrict our attention to state descriptions—that is, sentences of the form $\alpha_{i_1}(a_1) \wedge \alpha_{i_2}(a_2) \wedge \ldots \wedge \alpha_{i_n}(a_n)$—we get the following Weak Analogy Principle:

Weak Analogy Principle (WAP)
For atoms α_i, α_j, α_k, if $|\alpha_i - \alpha_j| < |\alpha_i - \alpha_k|$ then

$$w(\alpha_i(a_{n+1}) \mid \alpha_j(a_n) \wedge \theta) \geq w(\alpha_i(a_{n+1}) \mid \alpha_k(a_n) \wedge \theta)$$

for any *state description* $\theta \in QFSL$ containing only the constants a_1, $\ldots$, a_n.

We do know that there are many probability functions that will satisfy WAP. For example if we let w_1 be any probability function on L_1 satisfying Ex, we get a function w satisfying WAP on L_2 by defining

$$w\left(\bigwedge_{i=1}^{4} \alpha_i^{n_i}\right) = w_1(P_1^{n_1+n_2} \wedge \neg P_1^{n_3+n_4})w_1(P_2^{n_1+n_3} \wedge \neg P_2^{n_2+n_4})$$

It is an open problem to classify all functions satisfying WAP.

Finally, we note that this work only relates to a particular, simple kind of analogy and that more sophisticated distance functions, perhaps also depending on similarity between predicates, could be considered. Acceptable probability functions may well vary according to the distance function adopted.

References

Carnap, R. (1971). *Studies in inductive logic and probability* (Vol. 1). California: University of California Press.

Carnap, R. (1980). A basic system of inductive logic. In R. Jeffrey (Ed.), *Studies in inductive logic and probability* (Vol. 2, pp. 7–155). California: University of California Press.

Carnap, R., & Stegmüller, W. (1959). *Induktive Logik und Wahrscheinlichkeit*. Wien: Springer-Verlag.

Festa, R. (1996). Analogy and exchangeability in predictive inferences. *Erkenntnis, 45*, 229–252.

Hill, A., & Paris, J. (n.d.). An analogy principle in inductive logic. (Submitted to The Annals of Pure and Applied Logic)

Kuipers, T. (1984). Two types of inductive analogy by similarity. *Erkenntnis, 21*, 63–87.

Maher, P. (2001). Probabilities for multiple properties: The models of Hesse, Carnap and Kemeny. *Erkenntnis, 55*, 183–216.

Niiniluoto, I. (1981). Analogy and inductive logic. *Erkenntnis, 16*, 1–34.

Nix, C., & Paris, J. (2006). A continuum of inductive methods arising from a generalized principle of instantial relevance. *Journal of Philosophical Logic, 35*, 83–115.

Romeijn, J. (2006). Analogical predictions for explicit simmilarity. *Erkenntnis, 64*, 253–280.

Skyrms, B. (1993). Analogy by similarity in hyper-Carnapian inductive logic. In J. Earman, A. Janis, G. Massey, & N. Rescher (Eds.), *Philosophical problems of the internal and external worlds* (pp. 273–282). Pittsburgh: University of Pittsburgh Press.

Alexandra Hill
School of Mathematics, The University of Manchester
Manchester M13 9PL
e-mail: alex.hill@gmail.com
URL: http://personalpages.manchester.ac.uk/postgrad/alexandra.hill/

Jeff Paris
School of Mathematics, The University of Manchester
Manchester M13 9PL
e-mail: jeff.paris@manchester.ac.uk
URL: http://www.maths.manchester.ac.uk/~jeff/

The Confirmation of Singular Causal Statements by Carnap's Inductive Logic

Yusuke Kaneko

Abstract

The aim of this paper is to apply inductive logic to the field that, presumably, Carnap never expected: *legal causation*. Legal causation is expressible in the form of singular causal statements; but it is distinguished from the customary concept of scientific causation, because it is *subjective*. We try to express this subjectivity within the system of inductive logic. Further, by semantic complement, we compensate a defect found in our application, to be concrete, the impossibility of two-place predicates (for causal relationship) in inductive logic.

Keywords: Carnap, inductive logic

1 Problem of legal causation

What we call "singular causal statements" in this paper are the statements of the following kind:

(1) The X's parking on C Street caused the later traffic jam.

Hart and Honoré called this type of causation "legal causation" since it sometimes develops into a legal dispute (Hart & Honoré, 1985). We might as well call it "common-sense causation," considering its wider applications (cf. Hart & Honoré, 1985, p. 9). But in this paper, we uniformly call it "legal causation."

Legal causation is differentiated from scientific causation in that there are no support, such as *general laws* in Hempel and Oppenheim's schema[1]. Thus, here arises a problem: How can we *justify* legal causation?

---

[1] Cf. Hempel and Oppenheim (1948, especially pp. 138–140). Herein, Hempel and Oppenheim referred to Carnap's inductive logic as well (Hempel & Oppen-

The aim of this paper is to apply Carnap's inductive logic to this justification problem. Of course, I know that this is not in accord with its public image. Besides, legal causation is seemingly outside of Carnap's interest.

It was, rather, the theorists of *probabilistic causality* who treated this problem. For example, Patrick Suppes took up the statement very similar to (1), and showed interest in Hart and Honoré's work as well (Suppes, 1970, pp. 7–8). Nevertheless, I prefer Carnap's logic to probabilistic causality[2]. Why? To begin with, let me state the reasons.

2 Why I do not favor probabilistic causality

In appearance, probabilistic causality is suitable to analyze legal causation. But, in my view, it is still defective for the analysis.

Firstly, following probabilistic causality, we must reduce causal relationship to *conditional probability*, so that cause and effect are split into the two arguments of the probability function: P(*effect*, *cause*). Herein, the connection of cause and effect is considered in a mathematical way. But, I think, we intuit something real in causal connection, and it is not reducible to any mathematical relations. For instance, if you heard a dog barking when you tapped a desk, you will perhaps think: Your tapping the desk caused the barking. In this instance, we may say, you directly intuit the connection of the two events; and this intuition is concerned with something real, not reducible to mathematical relations. We must preserve this character of causation, but regrettably, probabilistic causality disregards it[3].

Secondly, most theories of probabilistic causality are practically comparative[4]. All they can do is showing the comparison like

---

heim, 1948, pp. 167f.). However, their reference were exclusively made for their theory of *systematic power*, that is, the power of the deductive systematization of a universal statement $\mathcal{T}$ over the data $\mathcal{K}$ in question. But it has little to do with our present interest.

[2] In this paper, the word "probabilistic causality" means "*a theory* of probabilistic causality" as well.

[3] As we shall see, in our application of inductive logic, this character is preserved. See (5)–(ii) below.

[4] According to Salmon, there were at least three theories of probabilistic causality so far (Salmon, 1980, p. 50): Good's *theory of causal network* (Salmon, 1980, pp. 51f.), Reichenbach's *causal theory of time* (Salmon, 1980, pp. 56f.), and Sup-

The Confirmation of Singular Causal Statements

$P(A_t, B_{t'}) \geq P(A_t)$. This analysis is, however, suitable for heuristics rather than for justification. Taking up Suppes' theory, for example, it narrows the class of *prima facie* causes (Suppes, 1970, p. 12) down to the *genuine* cause (Suppes, 1970, p. 24), *screening off spurious causes* (Suppes, 1970, p. 21, p. 24). This is an approach of heuristics. But we need a theory of justification now.

Even if probabilistic causality is regarded as a theory of justification, it will certainly not meet our requirements. Suppose, for example, we calculate the conditional probability of not being attacked, given that an individual was inoculated (cf. Suppes, 1970, pp. 12f.). For this calculation, probabilistic causality presupposes, in advance, the data that 749 people were not attacked within a total of 818 people; by comparison, 276 not attacked within 279 inoculated. And on the basis of this data, the calculation is made this way: $P(\textit{not attacked, the total}) = \frac{749}{818} \leq P(\textit{not attacked, inoculated}) = \frac{276}{279}$. This is how probabilistic causality concludes that inoculation is a (prima facie) cause of not being attacked.

However, our present object of study, legal causation, lacks very much this kind of objective data. Rather, the data used in it is, in most cases, *subjective*. And this subjectivity keeps away probabilistic causality from the analysis of legal causation.

3 Subjectivity of legal causation

But, why is legal causation so subjective? Where does the subjectivity come from? To clarify these points, let us consider the following scenario imaginable on the preceding example (=1).

(2) A policeman was searching for the cause of the traffic jam that occurred on C Street, which he thought was the cause of the traffic accident investigated. The accident happened at the moment when a certain driver spun into the opposite lane to avoid the traffic jam. He remembers that X had parked his car on this narrow street when he patrolled. A few weeks later, he judged: the cause of the traffic jam was the X's parking.[5]

---

pes' *probabilistic theory of causality* (Salmon, 1980, pp. 59f.). Reichenbach's and Suppes' were altogether comparative (Salmon, 1980, p. 60). Still, Good tried to make a quantitative theory; but it was ignored because of its "forbidding" style.

[5] There is a possibility to treat this problem in terms of *abnormality* (Hart &

Can we think this judgement objective? Presumably, the policeman observed C Street for a long time, and then, found that even short-term parking, such as X's, sufficiently caused a traffic jam. Again, based on this observation, he was convinced: the cause of the traffic jam was the X's parking (=1). Against this judgement, however, X can protest that on another street, such as A Street, even long-term parking rarely causes a traffic jam. In any case, we can say, the justification of legal causation is *subjective*.

This subjectivity of legal causation originates from its *context*. According to Hart and Honoré, the context of legal causation is much different from that of scientific causation (Hart & Honoré, 1985, p. 24). To take an example,

> (3) The sudden increase of traffic on C Street caused the later traffic jam.

We can classify this statement into scientific causation[6].

In some cases, we can be content with (3) as an explanation for the traffic jam. But in other cases, we cannot; we are tempted to ask an additional cause. This is because we cannot help attributing the harmful traffic jam to somebody else. Hart and Honoré called the contexts of this latter kind *attributive contexts* (Hart & Honoré, 1985, p. 24). And it was strictly distinguished from the former cases called *explanatory contexts* (Hart & Honoré, 1985, p. 24). It is this difference of contexts that differentiates legal causation from scientific causation, and makes legal causation *subjective*.

4 Similarity

In my opinion, inductive logic is suitable to express this subjectivity. Surely Carnap refuses this application. However, we can find a passage where he came close to our problem:

> (4) Suppose X owns a house whose value is $10,000, and he wonders whether to insure his house against fire. He will then

---

Honoré, 1985, pp. 37–40). But I leave this possibility aside in this paper.

[6]Define "traffic jam" as "vehicles slowing down to a specific speed." Then, "the increase of traffic" prevents vehicles from passing an intersection with a traffic light smoothly once. This suffices to cause a slowdown of vehicles, and so a traffic jam. A general law is of course available in this argument.

make his decision in view of the probability that his house will burn down during the next year. But, how can he predict it? He predicts it with regard to his knowledge e that contains information of previous experiences concerning *similar* houses.[7]

Here, Carnap admits that the *evidence* of inductive inference is gathered in terms of *similarity*[8]. This is true of (2) as well. In that situation, the policeman gathered the evidence in terms of similarity; concretely, he gathered the evidential events (parking) that all, similarly, occurred on C Street.

However, this choice is arbitrary. The policeman was certainly in a position to choose the street other than C. In fact, X can protest, against the policeman's judgement, that on another street, such as A, even long-term parking rarely causes a traffic jam.

The policeman may respond, against this objection of X's, that his evidence are all similar to the original case, the X's parking. But, against this response, X can further protest that the policeman's cognition of similarity is, after all, subjective.

5 Inductive logic

In my opinion, this subjectivity with a lax criterion of similarity is well expressed in inductive logic—this is the original aim of our inquiry. Let us then *design* a formal language, which constitutes, in a sense, a base of the following arguments[9].

(5) The Design of Language $\mathfrak{L}_3^2$

(i) "ε_1," "ε_2," and "ε_3" are the individual constants in $\mathfrak{L}_3^2$.

(ii) "__ is parking" and "__ causes a traffic jam" are the predicate constants in $\mathfrak{L}_3^2$.

---

[7] (Cf. Carnap, 1962, p. 256, p. 263). The sentences are modified by Kaneko.

[8] Similarity is dealt with in the argument of the *inference by analogy* as well (Carnap, 1962, pp. 207–208, pp. 567–571). But the similarity concept in our argument is about individuals. In contrast, the similarity in the inference by analogy is about properties.

[9] Carnap called this step a "classification of the signs" (Carnap, 1942, p. 24; Carnap, 1962, p. 65). Again, I want the readers to take note that "$\mathfrak{L}_N^\pi$" below means the language with N individual constants and π predicates (Carnap, 1962, p. 123).

$\mathcal{L}_3^2$ is a language of first-order logic including Davidson's logic of event (Davidson, 1967). Its informal explanation is as follows. "ε_3" is considered to be the X's parking on C Street. "ε_1" and "ε_2" are the evidential events gathered by the policeman. We can regard "ε_1" as the Y's parking on C street, and "ε_2" as the Z's parking on C street, for example. With this evidence, we can formulate the justification of the policeman's judgement (=1) as follows:

(6) $\mathfrak{c}^*((\varepsilon_3$ causes a traffic jam$),\{(\varepsilon_1$ is parking$) \wedge (\varepsilon_1$ causes a traffic jam$)\} \wedge \{(\varepsilon_2$ is parking$) \wedge (\varepsilon_2$ causes a traffic jam$)\} \wedge (\varepsilon_3$ is parking$))$

Here, "$\mathfrak{c}^*$" is a probability function peculiar to Carnap, which is called a $\mathfrak{c}$-*function*[10]. Its second argument, "$\{(\varepsilon_1$ is parking$) \wedge (\varepsilon_1$ causes a traffic jam$)\} \wedge ... \wedge (\varepsilon_3$ is parking$))$," expresses evidence, which is called an *individual distribution* (cf. (13) below).

How to assign concrete values to this formula is the core of the present study. Carnap's answer is this[11]:

(7) Let "s_M" be the number of individual constants of which molecular predicate M is predicated in evidence e. Further let "w_M" be the width of M, and "s" the number of individual constants observed up to then, "κ" the number of Q-predicates in $\mathcal{L}_N^\pi$. Then, the probability that M is predicated of the

---

[10]Strictly speaking, $\mathfrak{c}^*$ is no more than one option among many $\mathfrak{c}$-functions. In *Foundations*, Carnap narrowed all of possible $\mathfrak{c}$-functions down to this one (Carnap, 1962); its definition is (7) below.

[11]We think of this definition of $\mathfrak{c}^*$ as an expression of the subjectivity of inductive logic. But some might object that the subjectivity of inductive logic is adequately expressed in λ-system:

(†) $\mathfrak{c}_\lambda(M(\varepsilon_{s+1}), i) = \frac{s_M + \frac{w_M}{\kappa}\lambda}{s+\lambda}$ (Carnap, 1951, p. 33)

My answer to this objection is as follows. It is true that each person freely chooses λ's argument in (†), and we may perhaps attribute the subjectivity of inductive logic to that choice. However, the choice of λ is merely a choice of the inferential system (Laplace's system, Reichenbach's system, etc.); after the choice, however, everything works *objectively*. But our problem is why our inductive reasoning is *subjective* even after the choice. And my solution to this problem is: "Because Carnap's inductive logic is based on his possible world semantics. Considered in terms of this semantics, his inductive logic is likely regarded as the reflection of a personal view of the world." This opinion of mine is derivable only from *Foundation*. This is why now we stick to Carnap's former system, although the detailed argument is put off to another paper (Kaneko, 2010).

The Confirmation of Singular Causal Statements

next individual constants ε_{s+1} is calculated with the following formula:

$$\mathfrak{c}^*(M(\varepsilon_{s+1}), e) = \frac{s_M + w_M}{s + \kappa} \qquad \text{(Carnap, 1962, p. 568)}$$

We must follow up the unfamiliar words herein. For the explanation, we may as well divide the formula into two components: the logical factor $\frac{w_M}{\kappa}$ and the empirical factor $\frac{s_M}{s}$ (Carnap, 1962, p. 568). Firstly, the explanation of "κ" in $\frac{w_M}{\kappa}$ is provided.

(8) Only for abbreviation, we write "$P_1 \wedge P_2(e_1)$"[12] instead of "$P_1(e_1) \wedge P_2(e_1)$," for example, and call it a *molecular predicate expression*. Moreover, we can give the name "$M(e_1)$" to "$P_1 \wedge P_2(e_1)$," for example, and call it a *molecular predicate*. (Carnap, 1962, pp. 104–105)

Here, "P_1" and "P_2" are *primitive monadic predicates*, such as "__ is parking" and "__ causes a traffic jam" in $\mathfrak{L}_3^2$. In $\mathfrak{L}_3^2$, we can form four molecular predicates:

(9) $\forall e[Q_1(e) \longleftrightarrow (e \text{ is parking}) \wedge (e \text{ causes a traffic jam})]$
$\forall e[Q_2(e) \longleftrightarrow (e \text{ is parking}) \wedge \neg(e \text{ causes a traffic jam})]$
$\forall e[Q_3(e) \longleftrightarrow \neg(e \text{ is parking}) \wedge (e \text{ causes a traffic jam})]$
$\forall e[Q_4(e) \longleftrightarrow \neg(e \text{ is parking}) \wedge \neg(e \text{ causes a traffic jam})]$

These four molecular predicates "Q_1"~"Q_4" are called *Q-predicates*. Their formal definition is as follows:

(10) The molecular predicates defined in the following way are called *Q-predicates*.

$$\forall e[Q_i(e) \longleftrightarrow (\neg)P_1(e) \wedge \ldots \wedge (\neg)P_\pi(e)]$$
(Carnap, 1962, p. 125)

"P_1"~"P_π" are π primitive predicates in $\mathfrak{L}_N^\pi$. "$(\neg)$" stands for affirmation or negation. In general, there are 2^π Q-predicates in $\mathfrak{L}_N^\pi$ (Carnap, 1962, p. 125). "κ" in $\frac{w_M}{\kappa}$ expresses this number, 2^π.

Next, we take up the numerator of $\frac{w_M}{\kappa}$, that is, "w_M."

---

[12] This is also expressible as "$\lambda e[P_1(e) \wedge P_2(e)]$," using Church's lambda operator (Carnap, 1956, p. 3, p. 14).

(11) Any formula $\mathfrak{M}(e)$ in $\mathfrak{L}_N^\pi$ is expressed by a disjunction of Q-predicates as follows:

$$\forall e[\mathfrak{M}(e) \longleftrightarrow Q_{i1}(e) \vee Q_{i2}(e) \vee \ldots \vee Q_{iw}(e)] \quad \text{(Carnap, 1962)}[13]$$

By this theorem, we can substitute "$Q_1(\varepsilon_3) \vee Q_2(\varepsilon_3)$" for "($\varepsilon_3$ is parking)," for example. The number of Q-predicates which we substitute for formula $\mathfrak{M}$ is called the *width* of $\mathfrak{M}$ (Carnap, 1962, p. 127). It is marked with the second subscript "w" of the last disjunct in (11). "w_M" expresses it.

In this way, the logical factor is explained. Let us then proceed to the other factor, that is, the empirical factor.

(12) If molecular predicates $M_1, \ldots, M_p$ fulfill the following conditions, then they are called *forming a division* (Carnap, 1962, pp. 107–108).

(i) $\models {}^{14}\forall e[M_1(e) \vee \ldots \vee M_p(e)]$ \hfill (exhaustiveness)

(ii) For any M_i, M_j ($1 \leq i,j \leq p$), $\models \forall e \neg[M_i(e) \wedge M_j(e)]$ \hfill (exclusiveness)

(iii) For no M_i ($1 \leq i \leq p$), $\models \neg \exists e\, M_i(e)$
\hfill (M_i is not logically empty)

(13) The conjunction, in the following way, stating, over s individual constants and p molecular predicates forming a division, which predicate is predicated of which individual constant is called an *individual distribution*.

$$e_k = \lceil M_{k1}(\varepsilon_{j1}) \wedge M_{k2}(\varepsilon_{j2}) \wedge \ldots \wedge M_{ks}(\varepsilon_{js}) \rceil$$
\hfill (Carnap, 1962, p. 111)[15]

In $\mathfrak{L}_3^2$, one of the four Q-predicates in (9) occupies each position of "M_{k1}"~"M_{ks}."

"s" in the empirical factor expresses the number of individual constants in this individual distribution, and "s_M" expresses the number of individual constants in s that exemplify M, the predicate in question, which is one of "M_{k1}"~"M_{ks}."

---

[13] The proof was made in (Kaneko, 2010, (18)).
[14] "$\models$" means "logically true" though Carnap used "⊢."
[15] "⌈ ⌉" is Quine's quasi-quotes. But I place legibility prior to strictness in this paper.

6 Subjectivity

Now that we obtained the minimum knowledge of inductive logic, we can proceed to the calculation of (6), that is, the confirmation of legal causation.

(14) $\mathfrak{c}^*((\varepsilon_3 \text{ causes a traffic jam}), Q_1(\varepsilon_1) \wedge Q_1(\varepsilon_2) \wedge (\varepsilon_3 \text{ is parking}))$

from (9)

$$= \frac{\mathfrak{c}^*(Q_1(\varepsilon_3), Q_1(\varepsilon_1) \wedge Q_1(\varepsilon_2))}{\mathfrak{c}^*(Q_1(\varepsilon_3) \vee Q_2(\varepsilon_3), Q_1(\varepsilon_1) \wedge Q_1(\varepsilon_2))}$$

from def. of conditional probability and (11)

$$= \frac{\frac{2+1}{2+4}}{\frac{2+2}{2+4}}$$

from (7); note that both $Q_1(\varepsilon_1)$ and $Q_1(\varepsilon_2)$ exemplifies $Q_1(e) \vee Q_2(e)$

$$= \frac{3}{4}$$

In this way, we can trace the process of the policeman's judgement (=1). But I do not mean this is the actual process. My emphasis is, rather, on another point; that is, the policeman's conception over the evidence directly affected his reasoning. In other words, (14) is no more than the result of (5)[16]. To see this, let us consider another formation of language. Suppose, for example, X conceived the following formation in order to object against the policeman's judgement:

(15) The Design of Language $\mathfrak{L}^2_{3X}$

  (i) "ε_0," "ε_2," and "ε_3" are the individual constants in $\mathfrak{L}^2_{3X}$.

  (ii) "__ is parking" and "__ causes a traffic jam" are the predicate constants in $\mathfrak{L}^2_{3X}$.

Here, "ε_2" and "ε_3" are the same as in $\mathfrak{L}^2_3$. (X concedes in this respect.) But X removes "ε_1," and instead, puts "ε_0," which means the W's parking on A Street, for example. Thereby, he protests that ε_0 did not cause any traffic jam at all.

In this language, the probability of (6) is calculated as follows:

---

[16] Carnap also admitted that the design of language played an important role in inductive logic (Carnap, 1962, p. 54).

(16) $\mathfrak{c}^*((\varepsilon_3 \text{ causes a traffic jam }), Q_2(\varepsilon_0) \wedge Q_1(\varepsilon_2) \wedge (\varepsilon_3 \text{ is parking})) = \dfrac{1}{2}$

This is how the probability of the policeman's judgement is lowered below $\frac{3}{4}$ (=14).

This comparison of $\mathfrak{L}^2_{3X}$ with $\mathfrak{L}^2_3$ shows how influential the design of a language is in inductive logic. And the design is due to the person who wants to or refuses to confirm the legal causation in question. In this very respect, the subjectivity of inductive logic is brought to light.

7 The first criticism: on my subjective interpretation of inductive logic

In this way, Carnap's inductive logic gives a good framework to legal causation, which was advanced at the beginning of this paper. Let us then review this conclusion in the rest; that is, we scrutinize it from other viewpoints, especially from those of critics.

Firstly, let us take up our subjective interpretation of inductive logic. Some experts may say: "Carnap's inductive logic is concerned with the objective confirmation procedure in natural science. So your interpretation is besides the mark." However, it is relatively easy to respond to this criticism. As stated in Section 3, we have already entered an unexplored field, namely legal causation. And it is much different from the customary field of scientific causation. Thus, we may say, we have dealt with a completely new problem that Carnap never expected[17].

On the other hand, some experts on philosophy of probability may ask about the relationship between our subjective interpretation of inductive logic and Ramsey's *subjective theory*. As for this question, we can refer to Carnap's treatment of Ramsey's theory (Carnap, 1962, pp. 45–47). Therein, Carnap reduced Ramsey's subjective theory to his *logical theory*. Ramsey, in turn, admitted that probability theory is, in general, a branch of logic (Ramsey, 1926, p. 82 etc.).

---

[17] It is true that Canap had some ethical perspectives in his application of inductive logic to decision theory (Carnap, 1962, p. vii, pp. 252–279; Carnap, 1971); and, previously, I also followed this line (Kaneko, 2011). But now I came to think Hart and Honoré's distinction—between scientific causation and legal causation, or between explanatory context and attributive context—is more crucial for our argument.

Later, Carnap characterized his inductive logic as the pure and theoretical part of *normative decision theory* (Carnap, 1971, p. 26). Thereby, he regarded the agents following inductive logic as a kind of rational robot (Carnap, 1971, p. 17, p. 26). But, according to our analysis, even such agents cannot be perfect robots because, as we saw in Section 6, the source of their inference, the design of language, is far from mechanical objective procedures.

8 The second criticism: on my treatment of causal relation

The second criticism is against our awkward formulation of causation. In $\mathcal{L}_3^2$ (=5), we formulated causation in the following way:

(17) (ε_3 causes a traffic jam)

This formula is composed of one individual constant "ε_3" and a one-place predicate "__ causes a traffic jam." However, causation is nothing but causal *relationship*; so its formulation must be made with a two-place predicate like "__ causes __." Nevertheless, we have hitherto persisted in the one-place predicate.

To tell the truth, Carnap admitted two-place predicates in his system (Carnap, 1962, p. 114). But the problem is that he did not develop this idea any further[18].

In my opinion, it is impossible to develop the language with two-place predicates in inductive logic. One of the reasons is that Carnap confined his arguments to the language only with one-place predicates (Carnap, 1962, pp. 123f.). Therefore, all items, such as Q-predicate (cf. 10), were defined only by one-place predicates.

The theory that lies behind inductive logic is *combinatorics* (Carnap, 1962, pp. 156f.; Carnap, 1966, p. 23). If two-place predicates are introduced to inductive logic, the number of items, such as Q-predicate, will be extravagantly large[19].

---

[18] In "Meaning Postulates," Carnap touched on this problem once again. But it seems to me that he did not make any significant progress (Carnap, 1956, pp. 226–229).

[19] We can see this complexity even on one-place predicates. See (Carnap, 1962, p. 139)

However, this reason is not decisive. The true reason was the lack of theory. Carnap complained that there was no "theory for relations" in inductive logic, stating in parallel to the history of deductive logic (Carnap, 1962, pp. 123–124). Although he showed an optimistic attitude to this problem (Carnap, 1966, p. 33), such a theory for relations has not been developed yet. In my opinion, it is not necessary to invent such a theory for inductive logic. Instead, the lack of theory can be complemented by *semantics*. Let me elaborate on this idea below.

9 Carnap's semantics

As we shall see in the next section, however, our semantic complement is a kind of *model-theoretic semantics*. But the relationship between Carnap's system and model-theoretic semantics is not so clear. We must hence clarify the relationship between these two theories in advance.

Regarding this problem, two points are to be noted. Firstly, Carnap presumably did not know such model-theoretic semantics as we know today. Secondly, inductive logic is also classified into Carnap's semantics. Let us begin with this second point.

Carnap's semantics has two faces. One is the face obedient to Tarski's tradition: from the definition of truth (Tarski, 1933) to a theory of meaning (a truth-conditional theory of meaning)[20]. This face appears in Carnap's earlier studies of semantics (e.g. Carnap, 1942, pp. v–55).

The other face is L-semantics. This is the field for the explication of logical concepts, such as logical truth, logical consequence, and so on. The noteworthy here is the introduction of *state-descriptions*, Carnap's peculiar notion of *possible worlds*[21]. Based on this notion,

---

[20] But we must note: Tarski's concept of semantics is somewhat different from a theory of meaning (Tarski, 1944, p. 345). So we must take Davidson's program into consideration when we think about "Tarski's tradition" mentioned above (Davidson, 1962, p. 23).

[21] Let me define this notion for the subsequent arguments.

(†) The conjunctions introduced, as follows, by predicating one Q-predicate of each individual constants in $\mathfrak{L}_N^\pi$ are called state-descriptions:

$$\mathfrak{Z}_i = \ulcorner Q_{i1}(\varepsilon_1) \wedge Q_{i2}(\varepsilon_2) \wedge \ldots \wedge Q_{ic}(\varepsilon_N) \qquad \text{(Carnap, 1962, p. 116)}$$

Carnap thought, we could explicate the concept of probability as well (Carnap, 1942, pp. 96–97; Carnap, 1956, p. iii). That is why our present object of study, inductive logic, is classified into L-semantics[22].

In contrast, our semantic complement in the next section is a part of Tarski's tradition. To complicate matters further, it is stated in the form of *model-theoretic semantics*. It is true Tarski opened up the modern semantics[23]; but, even so, it still seems difficult to imagine model-theoretic semantics from Carnap's peculiar style.

As for this problem, Hintikka daringly severed the connection between these two theories (Hintikka, 1973). Presumably influenced by Church's criticism (Church, 1943), Carnap moved on to the study of intensional logic in *Meaning and Necessity*. Therein, according to Hintikka, model-theoretic semantics was closest at hand to Carnap's thought (Hintikka, 1973, p. 375)[24]. Nevertheless, Carnap did not lay his hand on it. This is because the model-theoretic semantics expected of him was Kripke-style possible world semantics (Hintikka, 1973, p. 374 etc.); Carnap adhered to his syntactic formulations of possible worlds—state-descriptions, so that he did not come up with Kripke-style semantics. That was why he failed in developing his theory to model-theoretic semantics (Hintikka, 1973, pp. 374–375, pp. 377–378)[25].

---

[22] Actually, Carnap regarded the c-function as a semantical function (Carnap, 1962, p. 164, p. 283, p. 522).

[23] As for the relationship between Tarski's argument and model-theoretic semantics, we can learn a lot from Raatikainen (2008). Therein, he indicated two points that differentiate Tarski's argument from model-theoretic semantics on the two parts of model-theoretic semantics: $\mathcal{M} = \langle \mathcal{D}, \mathcal{I} \rangle$. Regarding $\mathcal{D}$, he pointed out: Tarski's approach is possibly differentiated from the customary concept of domain in model-theoretic semantics, since Tarski seemingly considered only one fixed domain referred by an infinite sequence of objects (Raatikainen, 2008, p. 109). Regarding $\mathcal{I}$, he pointed out: $\mathcal{I}$ is possibly in contradiction with Tarski's commitment that he never presupposes any semantical concepts, since we can regard $\mathcal{I}$ as a semantical concept of designation (Raatikainen, 2008, pp. 112–113). But at the same time, in his article, a relief measure is also provided to reconcile these two theories (Raatikainen, 2008, p. 109, pp. 112–113).

[24] For example, like model-theoretic semantics, Carnap adopted *classes* as semantic values of predicate constants (Carnap, 1956, p. 19, p. 83). This idea is not found in his earlier system (Carnap, 1956, p. 166; Carnap, 1942, p. 18).

[25] Let us take an *individual concepts* as an example (Carnap, 1956, p. 41). This intensional object is considered to be a function that assigns one object in the discourse of universe $\mathcal{D}$ to each individual expression $\mathfrak{A}_i$ (e.g. "WBA boxing champion") with regard to a possible world w_j: $Intension(\mathfrak{A}_i, w_j) \in \mathcal{D}$ (Hintikka, 1973,

10 Semantic complement

Nevertheless, it was the same idea, state-descriptions, that enabled Carnap to complete his inductive logic. Should we then abandon model-theoretic semantics? My answer is, "No." If only L-semantics is narrowed down (not to include model-theoretic semantics in it), we may allocate model-theoretic semantics to the other face of Carnap's semantics: a truth conditional theory of meaning. This is how we may classify the following argument into the first face of Carnap's semantics, which gives a meaning to each expression of $\mathcal{L}_3^2$ in the manner differentiated from inductive logic.

Now then, let us embark on the semantic complement[26]. Therein, we aim at proving the following conditionalized T-sentence in context γ and in model $\mathcal{M}$:

(18) ((19) and (20)) $\Longrightarrow$ [{(17) is true in $\mathcal{M}$ in γ} $\longleftrightarrow$ (21)]

(17) is the sentence in question. (19), (20) and (21) are as follows:

(19) $\exists!e[(e$ is a traffic jam of C street$) \wedge (\text{T}(e) \subseteq \text{D-Term}(\gamma)) \wedge (\text{T}(e) < \langle now \rangle(\gamma))]$

(20) $\exists!e[\exists!x\{(e$ is parking of x by X$) \wedge (x$ is a car$)\} \wedge (e$ occurs on C street$) \wedge (\text{T}(e) \subseteq \text{D-Term}(\gamma)) \wedge (\text{T}(e) < \langle now \rangle(\gamma))]$

(21) $\iota e[\exists!x\{(e$ is parking of x by X$) \wedge (x$ is a car$)\} \wedge (e$ occurs on C street$) \wedge (\text{T}(e) \subseteq \text{D-Term}(\gamma)) \wedge (\text{T}(e) < \langle now \rangle(\gamma))]$

causes $\iota e[(e$ is a traffic jam of C street$) \wedge (\text{T}(e) \subseteq \text{D-Term}(\gamma)) \wedge (\text{T}(e) < \langle now \rangle(\gamma))]$

Here, "ι" is the iota operator, "T" expresses a function that assigns each event the time when it happens. "D-Term" expresses a function that assigns each context a *discourse term*[27]. "γ" is an individual constant that designates the context in situation (2). "$\langle now \rangle$" is Kaplan's

p. 376). To a certain extent, Carnap had this idea (Carnap, 1956, p. 181). Nevertheless, it was not developed; as just stated, his syntactical notion of possible worlds prevented him from introducing the primitive concept of w_j.

[26] It was already examined twice: in (Kaneko, 2009) and in (Kaneko, 2011).

[27] Tense expression is always concerned with a specific length of time. Suppose, for example, you ask, "Did he lock the door?" Then, it is not likely that you intended to ask whether he had ever locked the door. Like this example, when we use tense expression, we are supposed to have a specific length of time in mind. This is nothing but the discourse term stated in the text. See (Iida, 2002, pp. 338–339).

character (Kaplan, 1989, pp. 505f., p. 548)."$(t_1 < t_2)$" means that t_1 is earlier than t_2. It is to be noted that these are not Carnap's devices.

Let me explain (18) further. It says: "Provided that X parked his car on C Street in fact (=20) and there was a traffic jam on the street in fact (=19), then the expression '(ε_3 causes a traffic jam)' (=17) actually means, 'the X's parking caused the traffic jam' (=21=1)." Recall that (17) was composed of a one-place predicate "__ causes a traffic jam." On the contrary, (21) is composed of a two-place predicate "__ causes __." Hence, we can say, (17) actually means causal relationship represented by (21) if (18) is proved. This is the role of truth condition (18), which follows Tarski's tradition.

Hereafter, we make use of model-theoretic semantics to prove (18). Concretely, we introduce an intended model of $\mathcal{L}_3^2$: $\mathcal{M} = \langle \mathcal{D}, \mathcal{I} \rangle$. The statements below are parts of this model necessary for our proof:

(22) $\mathcal{I}(\text{``}\varepsilon_3\text{''}) = \iota e[\exists!x\{(e \text{ is parking of } x \text{ by X}) \wedge (x \text{ is a car})\} \wedge (e \text{ occurs on C street}) \wedge (\text{T}(e) \subseteq \text{D-Term}(\gamma)) \wedge (\text{T}(e) < \langle now \rangle(\gamma))]$

(23) $\mathcal{I}(\text{``\_\_ causes a traffic jam''}) = \{e_1 \mid \forall y[(e_1 \text{ occurs on } y) \rightarrow \exists!e_2((e_2 \text{ is a traffic jam of } y) \wedge (\text{T}(e_1) < \text{T}(e_2)) \wedge (\text{T}(e_1) \subseteq \text{D-Term}(\gamma)) \wedge (\text{T}(e_2) \subseteq \text{D-Term}(\gamma)) \wedge (\text{T}(e_1) < \langle now \rangle(\gamma)) \wedge (\text{T}(e_2) < \langle now \rangle(\gamma)) \wedge (e_1 \text{ causes } e_2))]\} = \{\mathcal{I}(\text{``}\varepsilon_1\text{''}), \mathcal{I}(\text{``}\varepsilon_2\text{''}), \mathcal{I}(\text{``}\varepsilon_3\text{''})\}$

(24) $\mathcal{I}(\text{``\_\_ is parking''}) = \{e \mid \exists s \exists x (e \text{ is parking of } x \text{ by } s)\} = \{\mathcal{I}(\text{``}\varepsilon_1\text{''}), \mathcal{I}(\text{``}\varepsilon_2\text{''}), \mathcal{I}(\text{``}\varepsilon_3\text{''})\}$

By the way, earlier, we made sure: the present semantics is classified into the first face of Carnap's semantics. Indeed, these statements are translatable into the language of that semantics under the name of "extensional neutral language M_e" (Cf. Carnap, 1956, pp. 168f.). Let us see the translation as well:

(25) "ε_3" designates the X's parking on C street.

(26) "__ causes a traffic jam" designates a cause of a traffic jam.

(27) "__ is parking" designates parking.

(25) corresponds to (22), (26) to (23), and (27) to (24), respectively.

The translation of (25) to (22) is not so problematic. If only we adopt the first-order language as M_e, we can somehow translate ordinary expression (25) into (22). Then, however, two points are to be noted: First, interpretation function $\mathcal{I}$ in (22) is the translation of the designation in (25)[28]. Second, the definite description appearing in (22) is not the expression of an individual concept (cf. note 25).

In contrast, the translation of (26) to (23) is more problematic. This is because the *class expressions*—besides, two—appear on the right side of (23). We can, however, make use of Carnap's notion of a *neutral entity* in this case (Carnap, 1956, pp. 153f.).

Let us regard "a cause of a traffic jam" in (26) as such a neutral entity. On the one hand, it is supposed to have an *intensional property*, which is expressed as the connotation on the first right side of (23). On the other hand, it is also supposed to have an extension, which is expressed as the extensional class expression on the second right side of (23).

In this way, we can interpret the two class expressions on the right side of (23) as the two aspects of one and the same neutral entity in (26). The same explanation is true of the translation of (27) to (24) as well. This is how we may say: the present model-theoretic semantics (semantic complement) is classified into the first face of Carnap's semantics.

Let us then return to the proof of (18). In this proof, firstly, we premise (19) and (20). These are factual statements. But premising factual statements in semantics is not question-begging. We can include empirical information in semantics, which was already shown in the extensional class expression in (23) and (24)[29].

On these premises (19) and (20), it suffices for the proof of (18) only to deduce its consequential part: the biconditional " '(ε_3 causes a traffic jam)' is true in $\mathcal{M}$ in $\gamma \longleftrightarrow$ (21)."

For the proof of this biconditional, we can refer to the empirical information stated in (23): $\mathcal{I}("\varepsilon_1") \in \mathcal{I}("\_\_ \text{ causes a traffic jam}")$. With this factual information, and from the customary definition of truth in model-theoretic semantics[30], we obtain the following:

(28) "(ε_1 causes a traffic jam)" is true in $\mathcal{M}$ in γ

Based on this fact, we can move on to the proof of $\Longleftarrow$ in the bicon-

---

[28]This is recognizable from the past controversy on Kripke semantics of the first-order modal logic.

[29]Cf. (Carnap, 1956, p. 70, pp. 163–164) and (Carnap, 1962, pp.126–127).

[30]"(ε_3 causes a traffic jam)" is true in $\mathcal{M}$ in $\gamma \longleftrightarrow \mathcal{I}("\varepsilon_1") \in \mathcal{I}("\_\_ \text{ causes a traffic jam}")$

ditional (in short, (28) ⟷ (21)). It is clear from the definition of material conditional that the problem for the proof of ⟸ is whether we can obtain (28) on the assumption of (21). But we have already obtained (28) above. So ⟸ holds.

The direction ⟹ is more problematic. On the assumption of (28), can we obtain (21)? By reference to (22)∼(24), firstly, we obtain the following statement from (28), based on the customary definition of truth (cf. note 30):

(29) $(\imath e[\ldots 22\ldots]$ occurs on C street$) \to \exists! e_2\{(e_2$ is a traffic jam of C street$) \wedge (T(\imath e[\ldots 22\ldots]) < T(e_2)) \wedge (T(\imath e[\ldots 22\ldots]) \subseteq$ D-Term$(\gamma)) \wedge (T(e_2) \subseteq$ D-Term$(\gamma)) \wedge (T(\imath e[\ldots 22\ldots]) < \langle now \rangle(\gamma)) \wedge (T(e_2) < \langle now \rangle(\gamma)) \wedge (\imath e[\ldots 22\ldots]$ causes $e_2))\}$[31]

Here we focus on the following theorem:

(30) $\exists! e A(e) \longleftrightarrow A(\imath e A(e))$[32]

We apply this theorem to (20) above; and from Conjunction Elimination, we obtain the following:

(31) $(\imath e[\ldots 22\ldots]$ occurs on C Street$)$

From Modus Ponens pertaining to (31) and (29), we obtain the following:

(32) $\exists! e_2\{(e_2$ is a traffic jam of C street$) \wedge (T(\imath e[\ldots 22\ldots]) < T(e_2)) \wedge (T(\imath e[\ldots 22\ldots]) \subseteq$ D-Term$(\gamma)) \wedge (T(e_2) \subseteq$ D-Term$(\gamma)) \wedge (T(\imath e[\ldots 22\ldots]) < \langle now \rangle(\gamma)) \wedge (T(e_2) < \langle now \rangle(\gamma)) \wedge (\imath e[\ldots 22\ldots]$ causes $e_2))\}$

Here, further, we focus on the following theorem:

(33) $(\exists! e A(e) \wedge \exists! e[A(e) \wedge B(e)]) \to \imath e A(e) = \imath e[A(e) \wedge B(e)]$[33]

We apply this theorem to (19) and (32); thereby, we obtain the following identity:

(34) $\imath e[\ldots 19\ldots] = \imath e[\ldots 32\ldots]$

---

[31]"$[\ldots 22\ldots]$" expresses the counterpart of (22). The same is true of the similar expressions below.
[32]The proof was made in (Kaneko, 2009, p. 50).
[33]The proof was made in (Kaneko, 2009, p. 51).

Again, we apply (30) to (32); and from Conjunction Elimination, we obtain this:

(35) $(\imath e[\ldots 22 \ldots]$ causes $\imath e[\ldots 32 \ldots])$

Finally, by the rule of substitution of identical things[34] with (35) and (34), we obtain (21) above. This is how $\Longrightarrow$ holds.

In this way, we could prove (18) in the intended model and in the proper context. Based on this, we may say, our awkward formulation (17) surely expresses the causal relationship between the X's parking and the traffic jam, which is nothing but the legal causation questioned at the beginning of this paper.

References

Carnap, R. (1942). *Introduction to semantics*. Harvard U.P.
Carnap, R. (1951). *The continuum of inductive methods*. Chicago U.P.
Carnap, R. (1956). *Meaning and necessity* (2nd ed.). Chicago U.P.
Carnap, R. (1962). *The logical foundations of probability* (2nd ed.). Chicago U.P.
Carnap, R. (1966). *Philosophical foundations of physics*. Basic Books.
Carnap, R. (1971). Inductive logic and rational decisions. In R. Carnap & R. Jeffrey (Eds.), *Studies in inductive logic and probability* (Vol. I). California U.P.
Church, A. (1943). Introduction to semantics. *The Philosophical Review*, *52*(3).
Davidson, D. (1962). Truth and meaning. In *Inquiries into truth and interpretation*. Oxford U.P.
Davidson, D. (1967). The logical form of action sentences. In *Essays on actions and events*. Oxford U.P.
Hart, H., & Honoré, T. (1985). *Causation in the law* (2nd ed.). Oxford U.P.
Hempel, C., & Oppenheim, P. (1948). Studies in the logic of explanation. *Philosophy of Science*, *15*(2).
Hintikka, J. (1973). Carnap's semantics in retrospect. *Synthese*, *25*(3/4).

---

[34] For any α and β, if $\alpha = \beta$ and $A(\alpha)$, we may infer $A(\beta)$.

Iida, T. (2002). *Gengo tetsugaku taizen IV* [The handbook of philosophy of language IV]. Keiso publishing company. (Japanese)

Kaneko, Y. (2009). [The phases of ethical judgements: From a motivistic point of view]. Unpublished doctoral dissertation, the University of Tokyo. (Japanese)

Kaneko, Y. (2010). Carnap's thought in inductive logic. (under refereeing)

Kaneko, Y. (2011). *Belief in causation: One application of Carnap's inductive logic.* (read at Philosophy of Science Colloquium in Institute Vienna Circle)

Kaplan, D. (1989). Demonstratives. In *Themes from Kaplan.* Oxford U.P.

Raatikainen, P. (2008). Truth, correspondence, models, and Tarski. In *Approacing truth.* Colledge Press.

Ramsey, F. (1926). Truth and probability. In *Philosophical papers: F.P. Ramsey.* Oxford U.P.

Salmon, W. (1980). Probabilistic causality. *Pacific Philosophical Quarterly, 61*.

Suppes, P. (1970). *A probabilistic theory of causality.* North-Holland Publishing Company.

Tarski, A. (1933). The concept of truth in formalized languages. In *Logic, semantics, mathematics.* Oxford U.P. (trans. by Woodger, J.H. (1956))

Tarski, A. (1944). The semantic conception of truth: and the foundation of semantics. *Philosophy and Phenomenological Research, 4*(3).

Yusuke Kaneko
The Faculty of Commerce, Meiji University, Japan
e-mail: kyaunueskuok_e@yahoo.co.jp

Logics of Fact and Fiction, Where Do Possible Worlds Belong?

John T Kearns

Abstract

This paper appeals to features of systems of illocutionary logic to understand and explain the difference between factual statements and illocutionary acts, on the one hand, and fictional statements and illocutionary acts, on the other. Systems of illocutionary logic have two levels, an ontic level which explores features determined by the truth conditions of statements, and an epistemic level which deals with relations of rational commitment among illocutionary acts performed with statements. The truth of a factual statement depends on that statement's relation to the real world, while the truth-in-fiction of a fictional statement depends on whether an ideal reader of a work of fiction, one who accepts or asserts the fictional statements governing the story or story world, is committed to accept or assert the fictional statement in question. The same illocutionary logical system is used for both factual and fictional statements and illocutionary acts, but that system is construed differently in the two cases.

Keywords: speech acts, fiction, illocutionary logic, logic of fiction

1 The logic of language acts

Illocutionary logic is the logic of *speech acts*, or *language acts*. A language act is a meaningful act performed by using an expression. Such an act can be spoken, or written, or thought, but we call them all speech acts as well as language acts.

Statements are language acts performed with sentences, they are language acts which are true or false, or, at least, which are evaluated in terms of truth and falsity. A number of semantic features of statements are determined by, or defined in terms of, truth conditions. For example, a statement is *analytically true* if its truth conditions can't fail to be satisfied, and it is *logically true* if it has a logical form whose truth conditions can't fail to be satisfied. A set or collection of statements *implies* a given statement if the statements in the collection have logical forms which can't be satisfied without satisfying the logical form of the given statement.

An *illocutionary act* is constituted by performing a sentential act with a certain force, like the force of an assertion, an apology, or a promise. The illocutionary act is the complete act that the language user intends to be performing when she performs the sentential act. An *assertive act* is an illocutionary act in which the sentential component is a statement, and in which the illocutionary force concerns the issue of whether the statement fits the world. Assertive acts include *assertions*, *denials*, and *positive* and *negative suppositions*.

An assertion is either an act of producing/performing a statement and accepting that statement as being or representing what is the case, or an act of producing a statement and reaffirming one's continued acceptance of the statement. A denial is an act which rules out the assertion of a statement, on account of the statement's failure to fit the world. To positively suppose a statement is to temporarily accept the statement, in order to explore the results of combining it with statements that are asserted or denied, and to negatively suppose a statement is to block or impede the temporary acceptance of the statement.

Systems of illocutionary logic have two levels, the *ontic* level, which is the level of statements and their truth conditions, and the *epistemic* level, which is the level of illocutionary acts and their conditions of rational commitment. An ontic level system of illocutionary logic is much the same as a standard system of logic. At the ontic level, it is customary to employ interpreting functions which determine sentences to be true or false, and to develop a deductive system for tracing truth-conditional connections linking statements and sets of statements. Implication or its converse, logical consequence, is often the focus of attention.

Asserting, denying, or supposing some statements *commits* a person to assert, deny, or suppose further statements. This *rational commitment* is conditional on a person taking an interest in the consequences of her illocutionary acts, and in the quite particular consequences of certain acts. For example, the person who asserts that today is Wednesday is committed by this assertion to accept that either today is Wednesday or it will snow tomorrow, but almost no one would have an interest in this consequence and actually consider and accept the disjunctive statement. However, it would be irrational to both accept that today is Wednesday, and refuse to grant that either today is Wednesday or it will snow tomorrow.

Commitment is either *immediate* or *mediate*. It is immediate commitment which is evident to the person who takes an interest, and immediate commitment which motivates a person to act. Mediate commitment is constituted by a chain of immediate commitments. A real-life argument is a *speech act argument* which moves from premisses which are assertions, denials, and suppositions to a conclusion which is also an act of one of these kinds. A *deductively correct argument* traces a chain of immediate commitments from the premiss acts to the conclusion act.

In an illocutionary logical theory, there is an artificial logical language which itself contains two levels. The first level contains *plain sentences* composed from non-logical expressions like unanalyzed atomic sentences, or predicates and individual constants, together with familiar logical expressions like connectives and quantifiers. Plain sentences represent statements. A first-level illocutionary theory deals only with plain sentences, while the second-level theory explores features of *completed sentences* and the illocutionary acts which they represent.

A completed sentence is obtained by prefixing a plain sentence with one of the following *illocutionary operators*:

$\vdash$ — the sign of assertion $\llcorner$ — the sign of positive supposition

$\dashv$ — the sign of denial $\neg$ — the sign of negative supposition

A second-level illocutionary theory is an epistemic theory for a person to use to represent and construct her own arguments. The semantic account for a second-level theory is designed for an ideal, or idealized, language user whom I call the *designated subject* and for whom I use feminine pronouns. The designated subject really knows what she thinks she knows, she has only true beliefs, and she doesn't forget what she knows and believes. We use an initial *commitment valuation* which awards the "value" + to the assertions and denials that constitute her explicit knowledge at a given time, or that constitute her explicit beliefs and disbeliefs at a given time, and this determines another commitment valuation, the *completion* of the initial valuation, which awards + to those assertions and denials which the initial valuation commits her to perform.

The second level deductive system contains (or represents) speech-act arguments which the designated subject constructs to explore the commitment consequences of her own assertions, denials, and suppositions.

2 Factual and fictional statements, and illocutionary acts

Ordinary statements, and ordinary assertions, denials, and suppositions are *factual*. A factual statement is one that is intended to be "measured" against the actual world. A factual assertion is an act of accepting a factual statement as "fitting" the actual world. Factual statements can be contrasted with *fictional* statements. A fictional statement is not one made by a character in a story, but is rather a statement about the characters and events in a story or "story world" that is made by a real person, by the author or a reader of/listener to the story. It is a factual statement that Al Capone committed many crimes, and a fictional statement that Sherlock Holmes solved many crimes.

Fictional *statements* are intended to be evaluated with respect to the story, or "story world," that they are talking about. Factual *illocutionary acts* are also distinguished from fictional ones. For example, a fictional assertion is an act of producing a fictional statement, and accepting or reaffirming one's acceptance of that statement as representing how things are in the story world.

The second level of an illocutionary logical theory, the epistemic level, focuses on assertive illocutionary acts and the completed sentences which represent them, the commitment conditions of the illocutionary acts, and features defined in terms of, or determined by, commitment conditions. Certain illocutionary acts *logically require* a further act if performing the first acts commits a person to perform the further act, and illocutionary acts are *incoherent* if they commit a person to assert and deny a single statement, or to positively and negatively suppose a given statement.

A person can assert, or accept, either a true factual statement or a false one, though if she accepts a false statement, she has made a mistake. With factual statements and factual illocutionary acts, what a person accepts or is committed to accept is a different matter from whether or not the statements involved are true. The truth or falsity of a factual statement depends on how the world is, not on who accepts the statement. With fictional statements, things are more involved. To understand the truth conditions of a fictional statement, we must consider two perspectives, an *internal* and an *external* perspective. The internal perspective is the perspective of a character in the story. From a character's perspective, there is an external world against which statements are measured. For a character, a statement is true if it fits that world, and false if it fails to fit the world. Every statement either fits or fails to fit the world.

The external perspective is the *reader*'s perspective. A sentence that a reader might use to make a fictional statement, if it were used by a character in the story, would be used to make (what is from the character's perspective) a factual statement. From the reader's perspective, there are different factors that contribute to truth and falsity in and for a story. One factor is what the author says or writes. Sherlock Holmes and Dr. Watson are related as Arthur Conan Doyle says they are, and they were involved in the various investigations, mysteries, and adventures that Conan Doyle tells about in the Sherlock Holmes stories. Another factor is genre conventions governing a type of story. In many science fiction stories, for example, the speed of light is not an upper bound on how fast characters can travel. Horror stories and vampire stories have conventions of their own. And then there is the default assumption that, unless the author or genre conventions say otherwise, things in the story "world" are pretty much the same as in the real world. Sherlock Holmes and Dr. Watson aren't (weren't) real people, but they lived in the real city of London, at the time of the real Queen Victoria. Victoria is Queen in Sherlock Holmes' world, and whatever she really did, she also did in his world. The Sherlock Holmes' stories don't talk about them, but Holmes must have had grandparents, great grandparents, etc. For Conan Doyle did not say otherwise.

The default assumption is not that the fictional world is like the real world in all respects that the author doesn't change. For it hardly makes sense to introduce the intricacies of quantum mechanics or relativity theory into Sherlock Holmes' world, or Hamlet's. The default assumption is that the fictional world is like the real world as the real world is/was currently known or believed to be. It might be a default assumption of some story that the world is flat, or that witches possess certain extraordinary powers and abilities.

Whether or not a fictional statement is true of the story, or story world, depends on the story, and not on the way our world really is. For there is no *extant* story world, there is no story world *there* against which fictional statements can be measured. So how is truth-in-fiction (from the reader's perspective) determined in an illocutionary theory? As it turns out, the two-level theories of illocutionary logic provide us with the resources we need to answer this question, and to explore the consequences of this answer. The short answer to the question is that, externally, a statement is true-in-fiction if an ideal reader of the story, one who knows the whole story (or stories), the genre conventions, and the default assumption, is committed to fictionally assert or accept that statement.

3 The two levels of an illocutionary logical theory

At the first level of an illocutionary logical theory, interpreting functions assign things to expressions, and determine each plain sentence to have either the value T or the value F. The epistemic-level, or second-level, semantic account is developed for a particular person, the one who is performing the illocutionary acts represented by completed sentences, and who is constructing speech-act arguments. Because logical theories have a normative aspect, the epistemic-level semantic account is developed for an ideal or idealized language user. At the epistemic level, an *initial* commitment valuation assigns + to those assertions and denials which the designated subject has performed and to which she remains committed, and this determines its *completion*, which assigns + to those assertions and denials that the designated subject is committed to perform on the basis of the knowledge or belief that is "registered" by the initial commitment valuation.

The same two-level system of illocutionary logic can be used for dealing with factual statements and illocutionary acts and for dealing with fictional statements and illocutionary acts, but the semantic account is construed differently in the two cases. (I should point out that in talking about what is true (or false) in fiction, I am considering one "story world" at a time. We consider our language to contain only sentences (representing statements) about this world, and evaluate them with respect to it.)

On the factual construal, the interpreting functions which award Ts and Fs to plain sentences are dispensing factual truth and falsity. On the fictional construal, the Ts and Fs that interpreting functions distribute are not truth-in-fiction and falsity-in-fiction for the reader. They are factual truth and falsity *from a character's perspective*. But we are readers, not characters in a story. Truth-in-fiction and falsity-in-fiction are determined by the commitment valuation which assigns + to the assertions and denials that the ideal reader is committed to perform on the basis of her knowledge of the story and of the genre conventions and default assumption. It is a default assumption for ordinary stories that classical logic holds in the story world. So excluded middle is a logical law in the story world. A sentence (statement) '$[A \vee \sim A]$' is logically true of the story world, and the assertion '$\vdash [A \vee \sim A]$' is "logically required."

A statement of excluded middle is also true-in-fiction from the reader's perspective. But from a character's perspective, the significance of the statement is that A is either true or false, while from the reader's perspective, the significance is not that every fictional statement is either true-in-fiction or false-in-fiction. One way that a reader might explain the significance of excluded middle is by saying that, for a character in the story, A is either true or false. A second way that the reader might explain its significance involves *ideal completions* of the story.

4 Stories are incomplete

It helps to understand this situation if we reflect on the fact that no author can tell a complete story. No matter how much he writes, there will be fictional statements A which are neither true-in-fiction nor false-in-fiction. But we can understand what it would be for a story or story world to be complete, even though no real-world author can tell such a story. For a given story or set of stories, there are any number of interpreting functions assigning T's and F's to atomic plain sentences, which agree with the story as told, and which complete the story. These are *ideal completions* of the story.

Every coherent story has infinitely many ideal completions. In the logical theory for fictional statements and illocutionary acts, the interpreting functions which award Ts and Fs to plain sentences, whose values agree with the story as told, represent ideal completions of the story. From the reader's perspective, that a disjunction '$[A \vee B]$' is true of the story world indicates that either A or B is true in every ideal completion of the story, even when neither A nor B is true in the story. So, from the reader's perspective, that a disjunction '$[A \vee B]$' is true-in-fiction is no sign that one of the disjuncts is true-in-fiction. The true disjunction doesn't *disjoin* the two statements A and B.

Because of this, it can make sense for a reader to introduce a new symbol for *genuine* disjunction *from her perspective*. If we use the symbol 'W' for "super" disjunction, or "reader's" disjunction, then a statement '$[A \, W \, B]$' will be true-in-fiction iff either A or B is true-in-fiction. This new connective has a modal significance. If we understand the box so that '$\Box A$' means that A's assertion follows (in the sense of commitment) from the ideal reader's knowledge of the story (including the genre conventions and default assumption), then '$\Box A$' has the significance that A is true of the story world, and we can define 'W' like this:

$$[A \, W \, B] =_{def} [\Box A \vee \Box B]$$

Once we introduce a modalized version of disjunction, it seems appropriate to introduce modalized versions of other connectives.

The connectives that strike me as most plausible for conjunction and negation are defined as follows:

$$[A \otimes B] =_{def} [\Box A \, \& \, \Box B] \quad \text{(Super Conjunction)}$$
$$\approx A =_{def} \Box \sim \Box A \quad \text{(Super Negation)}$$

For a super-negation '$\approx A$' to be true, it must be true-in-fiction that A isn't true-in-fiction. It is excessive to require that A be false in fiction. For if it is true-in-fiction that A isn't true in fiction, then supposing that A is true-in-fiction will lead to incoherence/contradiction.

The language (or sublanguage) that contains atomic sentences, compound sentences formed with super connectives, assertion, and super positive supposition:

$$\beth A =_{def} \neg \Box A$$

is characterized by an illocutionary version of Intuitionist logic. Super versions of denial and negative supposition can also be introduced, although they complicate the deductive system. It then requires slightly more work to show that the resulting system is a system of Intuitionist logic. Someone who regards mathematics as a special kind of fiction will adopt the reader's perspective, and should regard Intuitionist logic as the right logic for mathematics.

However, at present I am concerned with ordinary disjunction which indicates, from a character's perspective, that at least one disjunct is true, while from the reader's perspective, it only indicates that at least one disjunct is true in every ideal completion of the story. It is a feature of fictional discourse that an ordinary disjunction can be true of a story world when neither disjunct is true of that world. The disjunction is made true by the story teller, and not by some objective realm that the statement represents. Every story is essentially incomplete, and there are fictional statements that will never be determined to be either true or false.

Nobody thinks herself to be a character in a story, and living in a story world, although there are stories, including some movies, in which the characters do recognize themselves to be characters in the story. Someone who misunderstands some fictional discourse, and takes that discourse to be factual, is inadvertently adopting the perspective of a character. Her misunderstanding of the discourse might show up in her expecting every statement to be either true or false, in spite of the fact that there are true disjunctive statements for which there is nothing which determines either disjunct to be true.

5 Is mathematics fictional? Are possible worlds fictional?

Consider the proposal that mathematics is a special form of fiction. In the past, some mathematicians questioned the existence of negative numbers; later many, including Augustus DeMorgan, questioned the existence of imaginary, or complex, numbers. Eventually all mathematicians got "on board" with negative and complex numbers. Was this because they actually encountered these peculiar entities, or did they simply change their minds about the story they wanted to tell? Something similar happened more recently concerning the Axiom of Choice. Although it was once regarded with suspicion, it is now treated somewhat better.

Whether or not mathematics is a form of fiction, it is practically useful. We couldn't get along without it, either in science or in daily life. Whatever its status, an adequate philosophy of mathematics should explain that status, and also explain how we learn mathematics, investigate mathematics, and employ mathematics. In philosophy, talk of possible worlds, and employment of the conceptual apparatus associated with possible worlds, has become more and more prominent in the recent past. But what kind of talk is this? Is it factual or fictional? I think that much talk, perhaps all talk, about possible worlds displays the characteristics of fictional statements and illocutionary acts. In talking about possible worlds, perhaps one where kangaroos have no tails but get around on crutches, what is true in such a world is what we put in it when we describe that world. There is no way of telling whether the oldest kangaroo in a given world of tail-less kangaroos is, at some particular time, male or female, though it will be one or the other. There is no world *there* to settle things. All we have is a story.

If it turns out that mathematics is a special form of fiction, this doesn't detract from the importance, or the value, of mathematics. But if talk of possible worlds is a kind of fiction, then what good are possible worlds? If these worlds aren't really "out there," then how can they contribute to our understanding of metaphysical modalities? These worlds, and talk about them, will have an epistemic status, and perhaps have some relevance for epistemic modal logic, but they won't help us understand what things are really like.

John T Kearns
Department of Philosophy and Center for Cognitive Science
University at Buffalo, the State University of New York
Buffalo, New York 14260
e-mail: `kearns@buffalo.edu`

How to Build a Deontic Action Logic

Piotr Kulicki   Robert Trypuz*

Abstract

The aim of the paper is to point out the modelling choices that lead to different systems of deontic action logic. A kind of a roadmap is presented. On the one hand it can help the reader to find the deontic logic appropriate for an intended application relying on the information considering the way in which a deontic logic represents actions and how it characterises deontic properties in relation to (the representation of) actions. On the other hand it is a guideline how to build a deontic action logic which satisfies the desired properties.

Keywords: deontic logic, action theory, deontology, algebra of actions

Introduction

Most generally, deontic logic can be seen as a formal tool for analysing rational agent's behaviour in the context of systems of norms. Those norms can be of a different nature, moral, legal, technical are the main ones. Different systems of norms have different content and structure, and consequently, the correct schemas of reasoning about them may also differ. Thus, there may be many useful systems of deontic logic depending on their applications. Taking into account that fact we are interested in comparing the foundational principles of a group of deontic logics.

Research on deontic logic can be divided, from the technical point of view, into two main groups: in one of them deontic notions, such as permission, forbiddance or obligation are attributes of situations (in the language—propositions), in the other they are attributes of

---

*The project was funded by the National Science Center of Poland, decision number DEC-2011/01/D/HS1/04445.

actions (in the language—names). In the former one deontic notions are treated as a special case of modal operators and investigated with the use of tools of modal logic. So called *Standard Deontic Logic* is a logic of that type. In the present paper we are interested in the latter one—deontic action logic (DAL). The approach is used in the pioneering works of G. H. von Wright (1951) and J. Kalinowski (1953), the following papers (Fisher, 1961; Åquist, 1963) the paper of K. Segerberg (1982), which restored the "action approach" to deontic logic after the years of domination of the "propositional approach" and several papers developing the logics in question further. They include its presentation in a form of first-order theories (Lokhorst, 1996; Trypuz, 2011) and deontic logics of action built in connection with Propositional Dynamic Logic (PDL). In the latter class of systems two approaches can be distinguished. In one of them deontic operators are introduced with the use of dynamic operators and the notion of violation initiated in (Meyer, 1988), in the other one at least some of them are taken as primitive (McCarty, 1983; Meyden, 1996; Castro & Maibaum, 2009). Other recent work in the area include a system of deontic logic based on Kleene algebra (Prisacariu & Schneider, 2012).

The aim of the present paper is to point out the modelling choices that lead to different systems of DAL. We present a kind of a roadmap, which can help the reader to find a logic appropriate for an intended field of application on the basis of the way actions can be represented and how the deontic characteristics of actions relate to their representation. The decisions made in the construction of systems of deontic action logic are rarely explicitly stated by their authors, who usually treat their choices as obvious and their intuitions as the only possible ones. In the presentation we do not attempt to introduce new technical results, we concentrate on systems already defined in the literature, which is rich enough to provide a large number of examples. The paper is a continuation of our previous work form (Trypuz & Kulicki, 2009, 2010), but in contrast to it, we do not limit ourselves to DAL based on Boolean algebra, but also consider multi-valued deontic action logics and logics with sequences of actions.

We start from analysing the notion of action and the ways actions can be represented in deontic logic. In the following sections we study logics in which respectively one-step and multiple-step actions are considered.

1 Actions and their representation in the context of norms

At the beginning we will try to clarify the concept of action itself. In (Trypuz, 2008) several notions of actions are collected and analysed. Let us revoke some of the different definitions of action presented there:

- action is an event carried out by an agent (additionally one may require that the agent causes the event, has the intention to cause the event, has the reason for causing the event or that the event is constituted by a bodily movement of the agent);

- action is a bodily movement of an agent;

- action is an event, which is under control of an agent;

- action is a trying of an agent;

- action is a transition between states.

We shall not discuss which of those definitions is adequate. In the context of deontic logic it is enough to adopt the following features of actions present in or presupposed by most of the definitions:

- action is related to an agent, which is *responsible* for it;

- action has results, i.e. transforms the world from one state to another;

- actions can be combined.

Generally, in logic, valid inferences are based on the structure of formulas used in them. This applies also to DAL in which the structure of actions is crucial. In this case actions can be treated as composed from basic actions using several operators. Most common of them present in the papers on deontic logic are: parallel composition (doing two things at the same time), sequential composition (doing two things in sequel), free choice (performing an action arbitrarily chosen out of two) and negation of action (not doing something, refraining from doing something, doing something else).

Basic actions are least meaningful elements of agent's behaviour. They are chosen depending on the granularity of analysis and the type of considered agents, e.g.: "providing first aid", "paying $100 in cash", "making a bank transfer of $100", "turning left in the crossroads", "opening the window", "pressing the 'ok' button", "turning the switch on", "waving a hand".

As it was noticed already in (Wright, 1951) one can look at those actions as action individuals (particular actions performed by a concrete agent at a certain time in a certain place) or types of actions. In the latter case we use a name of an action as a description of many possible action individuals. It is important to be aware in which of those two meanings actions are used in a system of deontic logic, since one can build systems employing any of them, but they must not be mixed up.

Action operators, when applied to actions understood in one of those meanings, should return actions that can be understood in the same way. That causes some problems for actions understood as individuals. We can combine them using parallel and sequential composition but, when we apply action negation or free choice operator to them, we do not receive well defined individuals. Intuitively, not performing a particular action individual can be realised by many other action individuals. Doing one thing or another (the result of free choice between the two action individuals) is not a particular action individual as well. It can be realised by any of the two actions but it is not identical with any of them. Thus, if we want to use any of those two operators, we are forced to understand actions as types.

In DAL actions are arguments of deontic operators, usually permission, forbiddance and obligation. Any kind of norms considered in deontic logic by their nature can be applied to many action individuals. Action individuals are permitted, forbidden or obligatory not *per se*, but because they fulfil certain specification, e.g. a particular act of providing first aid to a person hurt in an accident is obligatory for a driver because it is always obligatory to provide first aid when an accident happens, not because the norm-giver predicted that this particular person at this particular place and time needs help.

When one understands action as types, this is achieved automatically. It is necessary, however, to be able to recognise actions in different situations as being of the same type. For example when we consider traffic regulations, my left turn at the crossroads in Hejnice last morning must be recognised as being of the same type as any left turn of any car at the same crossroads (or even any left turn at any crossroads).

When, on the other hand, we understand actions as individuals we have to categorise them, for example by using special predicates. We can then state that if a specific action α is an instant of providing first aid to a person hurt in an accident, then α is obligatory, for any action α. This approach leads us naturally to deontic logics formulated as first order theories. That gives one strong expressive power, but, on the other hand, causes high complexity problems. Moreover, DAL defined in such a way loses its specificity. Thus, in the rest of the paper we shall concentrate on logics in which actions are understood as types, which are more domain specific.

One can apply deontic characteristics to actions in two ways. The first way is to look only at the current situation of an agent and limit our judgements only to actions that may be performed in it. The other way is to consider also sequences of actions starting from the current situation. In the first one each situation can only be considered separately, in the other we may discuss a space of possible situations globally. From the technical point of view the distinction is between systems with or without sequential composition of actions. We shall refer to the systems employing those strategies respectively as one-step and multi-step deontic action logics.

We shall use the following formal notation: $a, b, \ldots$ (possibly with subscripts) will stand for basic actions; $\alpha, \beta, \ldots$ will represent any actions; $\mathbf{0}$ and $\mathbf{1}$ will stand for impossible action and universal action (doing anything) respectively; *overline* will be used for unary operator of action negation; $\sqcap, ;, \sqcup$ will represent respectively binary operators of parallel and sequential compositions of actions and free choice between actions; P, F and O will be used respectively for permitted, forbidden and obligatory predicates.

The following definition of a well formed formula of DAL is a usual routine:

$$\phi ::= \perp \mid \mathsf{F}(\alpha) \mid \mathsf{P}(\alpha) \mid \mathsf{O}(\alpha) \mid \neg\phi \mid \phi \to \phi$$

$$\alpha ::= a \mid \mathbf{0} \mid \mathbf{1} \mid \overline{\alpha} \mid \alpha \sqcap \alpha \mid \alpha \sqcup \alpha \mid \alpha;\alpha$$

2 One-step systems

This section is divided into two parts. In the first one we consider systems of DAL in which deontic value of complex action is a function of deontic values of basic actions that are used to define that complex actions. In those systems deontic matrices can be used. In the other, systems of DAL in which more complex attribution of deontic notion to complex actions are considered.

2.1 Multivalued deontic action logic

The basic idea of multivalued DAL is to mark each basic action by its deontic value, calculate deontic value of complex actions on that basis, and finally, truth value of a deontic formula.

Historically the first and also the simplest system of this kind of Kalinowski and Fisher[1] is defined using three deontic values: good (g), neutral (n) and bad (b) by the following matrices:

α	$P(\alpha)$	$O(\alpha)$	$F(\alpha)$
b	0	0	1
n	1	0	0
g	1	1	0

$\sqcap$	b	n	g	$\overline{\alpha}$
b	b	b	b	g
n	b	n	g	n
g	b	g	g	b

Intuitively one may understand the first matrix as stating that good and neutral actions are permitted and bad ones are not, good actions are obligatory and neutral and bad ones are not, and bad actions are forbidden and good and neutral ones are not. The second matrix informs us how to obtain a deontic value of negation and parallel composition of two actions. Free choice operator is understood as a de Morgan dual of parallel composition.

The least obvious is the deontic characterisation of *a parallel composition of a good and a bad action*. The matrix states that such an action is always bad. That may be unacceptable in some contexts. The problem can be solved within the multi-valued approach, since the system is not a single possible DAL defined by means of matrices. One of the alternative solutions[2] is based on the lattice inspired by N. Belnap's lattice of truth and information (Belnap, 1977).

---

[1] Kalinowski introduced a general idea and the matrix for negation, Fisher added matrices for $\sqcap$ and $\sqcup$.

[2] The formal details of the system sketched here and other possible multivalued DAL are the subject of our ongoing research.

How to Build a Deontic Action Logic

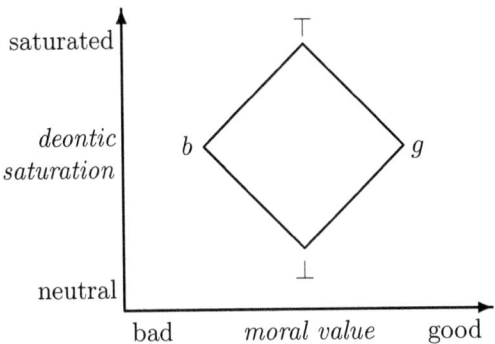

Figure 1: A bilattice of moral value and deontic saturation

In the resulting system four deontic values are used: good (g) and bad (b), as in the previous system, and two new ones: bottom ($\bot$) and top ($\top$). Bottom value is understood as neutral and top as oversaturated, which is applied to action defined as a combination of good and bad elements. The interpretation of deontic operators is given by the following matrix.

α	$P(\alpha)$	$O(\alpha)$	$F(\alpha)$
b	0	0	1
$\bot$	1	0	0
$\top$	1	0	0
g	1	1	0

One may notice that both $\top$ and $\bot$ are treated as n in the system of Kalinowski and Fisher.

The parallel composition and free choice are interpreted respectively as supremum and infimum of the lattice from Figure 1. We can also present those operators along with action negation in the following matrices:

⊓	b	⊥	⊤	g	ᾱ
b	b	b	⊤	⊤	g
⊥	b	⊥	⊤	g	⊥
⊤	⊤	⊤	⊤	⊤	⊤
g	⊤	g	⊤	g	b

⊔	b	⊥	⊤	g
b	b	⊥	b	⊥
⊥	⊥	⊥	⊥	⊥
⊤	b	⊥	⊤	g
g	⊥	⊥	g	g

In contrast to the system of Kalinowski and Fisher a combination of good and bad actions is now neutral (is neither forbidden nor obligatory).

In both presented multivalued systems deontic operators of permission, forbiddance and obligation are interdefinable in the sense that any one of them is sufficient to define the remaining ones.

2.2 Deontic action logics based on Boolean algebra

More sophisticated considerations on deontic value of complex actions can be performed with the use of techniques of Boolean algebra of actions[3].

The first problem to discuss is whether a set of basic actions should be finite. In (Segerberg, 1982) it is not, while in (Castro & Maibaum, 2009) it is. Using finite sets makes the algebra atomic, which is pleasing from the technical point of view. Moreover, it seems to be a natural restriction, since actions are understood as types and it is reasonable to consider a finite number of action types. If we recognise infinitely many types of basic actions we can always group them into a finite set of categories.

In systems of DAL based on Boolean algebra known from the literature, in contrast to multi-valued DAL, permission and forbiddance are not inter-definable. Permitted and forbidden actions are understood in a so called strong sense as permitted and forbidden in any context (in combination with any other possible actions). A basic system of that kind can be defined by the following axioms introduced in (Segerberg, 1982):

$$P(\alpha \sqcup \beta) \equiv P(\alpha) \wedge P(\beta) \tag{1}$$

$$F(\alpha \sqcup \beta) \equiv F(\alpha) \wedge F(\beta) \tag{2}$$

$$\alpha = \mathbf{0} \equiv F(\alpha) \wedge P(\alpha) \tag{3}$$

---

[3]The structures of deontic values of the systems from the previous section can also be treated as algebras, but the negation there is not the negation of Boolean algebra and the number of elements is limited to 3 or 4.

How to Build a Deontic Action Logic

That opens an opportunity to discuss problems of openness and closedness of deontic action logic. For which class of actions an action is either permitted or forbidden? Basic actions and atomic actions are candidates. The following formulas expressing the discussed properties can be considered as different axioms of DAL (Act_0 is a set of basic actions) that can be added to a basic open system:

$$F(a_i) \vee P(a_i), \text{ for } a_i \in Act_0 \tag{4}$$

$$P(\overline{a_1} \sqcap \cdots \sqcap \overline{a_n}) \vee F(\overline{a_1} \sqcap \cdots \sqcap \overline{a_n}), \text{ where } \{a_1, \ldots, a_n\} = Act_0 \tag{5}$$

$$(a_1 \sqcup \cdots \sqcup a_n) = \mathbf{1} \tag{6}$$

$$F(\delta) \vee P(\delta), \text{ for } \delta \text{ being an atom of algebra.} \tag{7}$$

Relation between systems resulting from using the above formulas as axioms are illustrated in Figure 2.[4]

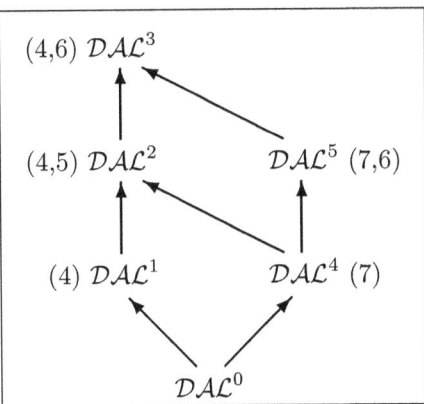

Figure 2: A structure of $\mathcal{DAL}^n$ systems

Moreover, we may introduce obligation operators in different manners. In (Trypuz & Kulicki, 2010) we discussed several notions of obligation defined on the basis of permission and forbiddance. Out of them most turned out not to fulfil the basic intuitions about obligation. The one

---

[4] Proofs of the results concerning the relations between the systems and semantic presentation of them one can find in (Trypuz & Kulicki, 2009).

worth mentioning is defined as follows:

$$\mathsf{O}(\alpha) \triangleq \mathsf{P}(\alpha) \wedge \mathsf{F}(\overline{\alpha}) \qquad (8)$$

The problem is that Definition 8 allows only one action to be obligatory, that is:

$$\mathsf{O}(\alpha) \wedge \mathsf{O}(\beta) \to \alpha = \beta \qquad (9)$$

Alternatively to Definition 8, obligation can be introduced independently from permission and forbiddance as a primitive notion introduced with the following axioms:

$$\mathsf{O}(\alpha) \wedge \mathsf{O}(\beta) \to \mathsf{O}(\alpha \sqcap \beta) \qquad (10)$$
$$\neg \mathsf{O}(\mathbf{0}) \qquad (11)$$
$$\mathsf{O}(\alpha) \to \mathsf{P}(\alpha) \qquad (12)$$

3 Multi-step systems

Propositional dynamic logic (PDL) is a modal logic created to model behaviours of computer programs (Harel, 1984). It was soon noticed by philosophers that it can also be applied *per analogiam* to represent human actions (Segerberg, 1980). Actions in PDL are interpreted as transitions between states. In the language of this logic we also find interesting action-forming operators, which allow for expressing free choice, parallel execution, refraining, sequential processing and action repetition.

As we have mentioned in the introduction, there are two main approaches to combining deontic logic with PDL. In one of them deontic operators are introduced in Andersonian style, with the use of dynamic operators and the notion of violation (Meyer, 1988), in the other one at least some of them are taken as primitive (McCarty, 1983; Meyden, 1996; Castro & Maibaum, 2009).

Dependening on the particular type of deontic logic built in connection with PDL, the idea of modelling norms slightly differs. Some of them focus on the actions/transitions—describing them as legal (green) or illegal (red)—whereas the other ones take into account the states which actions/transitions lead to—describing them as desired or undesired (by the norm-giver).

One of the basic problems of DAL with sequential composition is the introduction of action negation. The notion of action negation (or complement) is problematic in action theory in general (see e.g. Segerberg, 1996). In the context of deontic action logic with sequential composition the problem was studied in (Broersen, 2004). The problem has two levels: intuitive and technical. The intuitive level is how to understand a negation of action when actions consists of multiple steps. The technical level is that the calculi containing sequential composition and complement of relations are undecidable and difficult to axiomatise.

The most straightforward way to understand negation of action α is by analogy to relation complement. One takes the set of all possible complex actions available for an agent, let us call it R, and action $\overline{\alpha}$ is understood as $R \backslash \alpha$ (R can also be understood as the set $S \times S$, where S is the set of all the possible states of the world). This is, however, unintuitive for actions, because gives us much more then actions alternative to α at a given situation. Moreover, using such a notion of negation leads to an undecidable system.

Broersen (2004) introduces another notion of negation by restricting the set R to actions available for an agent in a current state. In terms of state space he uses only reachable states. Depending on an accessibility relation on states, he receives different members of a family of negation operators.

Another notion of action negation comes from (Wansing, 2005). In that approach the standard de Morgan laws for negation, parallel composition and free choice, and the following law:

$$\overline{\alpha; \beta} = \overline{\alpha}; \overline{\beta} \qquad (13)$$

holds. That allows us to push the negation down to the basic actions. For that reason this notion is pleasing from the technical point of view. However (13) is not intuitively clear and there is no clear connection between a basic action and its negation.

Yet another notion of negation is used in (Meyer, 1988; Prisacariu & Schneider, 2012). In that notion the basic idea is that refraining from performing a sequence of actions is a free choice between not doing the first of them and a sequence of doing the first of them and not doing the rest. Formally:

$$\overline{\alpha; \beta} = \overline{\alpha} \sqcup \alpha; \overline{\beta} \qquad (14)$$

The difference is that in the latter paper this notion is introduced as a defined operation restricted to actions in canonical form. That restriction makes the system decidable.

The other problem is how to relate deontic predicates with sequences of actions. Should obligatory actions be all obligatory instantly? Formally such a constraint can be expressed by formula (10). An obligation operator respecting (10) is called in (Prisacariu & Schneider, 2012) a natural obligation. Alternatively, we may think of its weaker version in which two obligatory action can be performed simultaneously or consecutively, formally:

$$O(\alpha) \wedge O(\beta) \rightarrow O(\alpha \sqcap \beta) \vee O(\alpha; \beta) \vee O(\beta; \alpha) \qquad (15)$$

Another problem is when the sequence of two actions is forbidden. In (Prisacariu & Schneider, 2012) it is so when both actions are forbidden. Alternatively one may consider a sequence as forbidden, when one of its elements is forbidden.

Conclusions

In the paper we point out that the crucial factors that influence the construction of any particular system of deontic action logic are the way actions are represented and the way deontic operators are attached to elements of that representation.

References

Åquist, L. (1963). Postulate sets and decision procedures for some systems of deontic logic. *Theoria, 29*, 154–175.

Belnap, N. (1977). A useful four-valued logic. In G. Epstein & J. M. Dunn (Eds.), *Modern uses of multiple-valued logic* (p. 7–37). Dordrecht: Reidel.

Broersen, J. (2004). Action negation and alternative reductions for dynamic deontic logics. *Journal of Applied Logic, 2*(1), 153–168.

Castro, P. F., & Maibaum, T. S. E. (2009). Deontic action logic, atomic boolean algebras and fault-tolerance. *Journal of Applied Logic, 7*(4), 441–466.

Fisher, M. (1961). A three-valued calculus for deontic logic. *Theoria*, *27*, 107–118.

Harel, D. (1984). Dynamic logic. In D. Gabbay & F. Guenthner (Eds.), *Handbook of philosophical logic, volume II: Extensions of classical logic* (Vol. 165, pp. 497–604). Dordrecht.

Kalinowski, J. (1953). Theorie des propositions normatives. *Studia Logica*, *1*, 147–182.

Lokhorst, C. G.-J. (1996). Reasoning about actions and obligations in first-order logic. *Studia Logica*, *57*(1), 221–237.

McCarty, L. T. (1983). Permissions and obligations. In *Proccedings of IJCAI-83* (pp. 287–294).

Meyden, R. V. D. (1996). The dynamic logic of permission. *Journal of Logic and Computation*, *6*, 465–479.

Meyer, J. J. C. (1988). A different approach to deontic logic: Deontic logic viewed as a variant of dynamic logic. *Notre Dame Journal of Formal Logic*, *1*, 109–136.

Prisacariu, C., & Schneider, G. (2012). A dynamic deontic logic for complex contracts. *The Journal of Logic and Algebraic Programming*, *81*, to appear.

Segerberg, K. (1980). Applying modal logic. *Studia Logica*, *39*(2–3).

Segerberg, K. (1982). A deontic logic of action. *Studia Logica*, *41*(2-3).

Segerberg, K. (1996). To do and not to do. In J. Copeland (Ed.), *Logic and reality: Essays on the legacy of arthur prior* (p. 301–313). Oxford University Press.

Trypuz, R. (2008). *Formal ontology of action: A unifying approach.* Wydawnictwo KUL.

Trypuz, R. (2011). Simple theory of norm and action. In A. Brożek, J. Jadacki, & B. Žarnić (Eds.), *Theory of imperatives from different points of view* (pp. 120–136). Wydawnictwo Naukowe Semper.

Trypuz, R., & Kulicki, P. (2009). A systematics of deontic action logics based on boolean algebra. *Logic and Logical Philosophy*, *18*(3–4), 253–270.

Trypuz, R., & Kulicki, P. (2010). Towards metalogical systematisation of deontic action logics based on Boolean algebra. In G. Governatori & G. Sartor (Eds.), *Deontic logic in computer science (deon 2010)* (pp. 132–147). Springer.

Wansing, H. (2005). On the negation of action types: Constructive concurrent pdl. In *Logic methodology and philosophy of science. Proceedings of the twelfth international congress* (p. 207–225). London: King's College Publications.

Wright, G. H. von (1951). Deontic logic. *Mind*, *LX*(237), 1–15.

Piotr Kulicki and Robert Trypuz
The John Paul II Catholic University of Lublin
Al. Raclawickie 14, Lublin, Poland,
e-mail: {kulicki,trypuz}@kul.pl

The Matter of Objects

Henry Laycock

Abstract

In formal, logico-semantic terms, the concept of an *object, individual* or *thing* is widely thought to be exhaustive. Whatever we may say or think—it is supposed—our thought or talk is always thought or talk of objects. Here, I briefly sketch out and defend a rival view: like that of a *property*, the concept of an object is best understood in terms of that which it excludes. Object-concepts correspond to *count* nouns; but count nouns are just one of two great categories of nouns, the other being that of *non-count* nouns. Both count nouns, and the concept of an object, are inseparable from reference and identity—but non-count nouns are not; and they are not a vehicle for thought or talk of objects in the first place.

Keywords: Frege, Russell, Quine, Strawson, count nouns, mass nouns, quantification

1

The category of *objects, individuals* or *things*—the concept of *an* object, individual or thing—is widely thought to be exhaustive. Whatever we may say or think, it is supposed, our thought or talk is always thought or talk of objects. In ontic terms, this world—"our" world—cannot be other than a world of objects. Alluding to the intimacy of the link between the concept of an object and the notion of the *subject* of a thought or proposition, Russell (1903) famously writes

> Whatever may be an object of thought, or may occur in any true or false proposition, or can be counted as one, I call a term. This, then, is the widest word in the philo-

sophical vocabulary. I shall use as synonymous with it the words unit, individual and entity. (Russell, 1903, p. 43)[1]

There is quite simply *nothing*, on this view, that the concept of an object might exclude; but this view, I will urge, is ultimately no more than a prejudice. Admittedly, it is a very special prejudice—the doctrine is an ancient one, and embraced by eminent authorities.[2] But yet it can be challenged on semantic grounds; and in the very work in which he advocates the view, Russell indicates concern as to its truth.[3] Against the doctrine that the concept of an object excludes nothing, here I mean to argue that the concept is best understood *precisely* via that which it excludes.

Like Frege, Russell does not, of course, believe that *all* propositions are of the traditional {individual object / grammatical subject} / {general concept / grammatical predicate} form. The rejection of that view is a key feature of Frege's groundbreaking conception of quantification—a feature taken up by Russell, and brilliantly reflected in the Theory of Descriptions. Rather, on the model of language shared by both logicians, sentences are divided into two broad categories. On the one hand are *referential* sentences, which do exemplify the object / general concept model, and on the other hand are *quantified* or non-referential sentences, constructed on the basis of the former group. On the one hand are sentences whose basic form is just "Fa"; on the other hand there is "$(\exists x)(Fx)$".

2

It is difficult to find fault with Russell's assertion that whatever may be counted as an object, in this very general formal sense, is

---

[1] By "unit", clearly, Russell does not mean "unit of __", as in, e.g., "unit of *length*"; he means whatever can in some sense stand alone as a unit, whether that be a cat, a number, or a knot.

[2] A number of authorities are cited in (Laycock, 2006). See especially ch. 3.

[3] Russell (1903) is clearly very worried both by non-singular predication, as exemplified in "Brown and Jones are two of Miss Smith's suitors", and indeed by the limitations of the doctrine that "whatever is, is one". Against this doctrine, he remarks that "whatever are, are many"; and goes on—in oddly paradoxical terms—to insist that a "plurality of terms is not the logical subject when a number is asserted of it; such propositions have not one subject, but many subjects" (Russell, 1903, p. 69, fn.). The matter is rehearsed in detail in (Laycock, 2006, ch. 2).

thereby to be counted *as one*. Yet, if it is an essential feature of the object-concept that objects must be capable of being counted and distinguished, then instantiated thus in countables, the object-concept seems plainly to be limited in scope. The semantic category in which the concept is linguistically represented is that of *count nouns* or *CNs*; such nouns are precisely words for discrete objects, individuals or things. But the fact is that the category of CNs is not the only category of nouns. There exists a major and distinct semantic category of *non-count* nouns or *NCNs*—words like "smoke", "rain", "water", "air", and "gold". And much as concrete CNs are words for concrete individuals or *things*, so concrete NCNs are words for *stuff* or matter. As categories of nouns, concrete CNs and NCNs constitute two great semantic categories—by definition mutually exclusive, and more or less exhaustive—categories which roughly correspond to the *ontic* categories of stuff and things, albeit suitably defined.

Now it is a direct implication of the object / concept model, that *to be* is for a concept or kind to be *instantiated*—to have, that is, at least one instance. And it is not surprising that if questions are raised about those concepts or kinds which are concepts or kinds of *matter*, just this paradigm will be invoked. Much as for "ordinary" concrete objects, for kinds of matter to exist will be for those kinds also to have instances. The thought seems *prima facie* plausible: the instances in question would seem to be such entities or objects as *the milk* in this glass and *the milk* in that, and likewise, for any other kind of stuff. Indeed, simply insofar as we are able to *distinguish* this milk from that, it hardly seems outrageous to suppose we are *thereby* dealing with distinct objects. Instances of kinds of stuff, no less than instances of kinds of organisms, will figure as values of variables in the formal representation of natural-language sentences.

But plausible though this thought seems to be, it has only to be questioned to be doubted. For CNs are semantically *either* singular or plural: singularity and plurality are the twin semantic sub-categories which jointly exhaust this category of nouns. It appears therefore to follow that the category of NCNs can be *neither* singular nor plural—a fact which is obscured if, instead of *non-count noun*, the more common appellation *mass noun* is employed. And given this dichotomy, two fundamental propositions follow directly.[4] On the one hand, NCNs

---

[4]These semantic points are argued at some length in (Laycock, 2006). See in

	1. Singular ("one")	2. Non-singular ("not-one")
3. Plural ("many")		"Objects"
4. Non-plural ("not-many")	"Object"	"Stuff"

Table 1

must share with plural nouns the distinction of being semantically *non-singular*.[5] On the other hand, NCNs must share with singular nouns the distinction of being semantically *non-plural*. This then yields the following very simple taxonomy, see Table 1.[6]

For the object-category, as Russell observes, objecthood is not distinct from *unity*. Unity is the constitutive, objective principle of the category—*an* object, as in $\{4,1\}$, just is an ontic *unit*; equally, *objects*, as in $\{3,2\}$, are, quite simply, ontic *units*. But *matter*, by contrast, as in $\{2,4\}$ has no principle of unity, hence cannot show itself as *multiplicity*. Rather, it has a distinct principle of *quantity* or *amount*—things are either one or many; stuff is merely more or less. On this account, the categories of matter and of objects are quite fundamentally distinct, and it is an implication of the non-singularity of $\{2,4\}$ that there can be no *instances* of matter-concepts. It follows that we cannot, in general, understand existence claims via the "instantiation" of concepts—at least, we cannot understand it in these terms, *if* "to be instantiated" means "to have at least one instance". Words like "water", "air", and "gold" can have no concrete instances. But what exactly does this negative claim amount to?

---

particular chs. 1, 3 and 4.

[5] Indeed, singularism notwithstanding, the semantic kinship between NCNs and plural nouns is these days widely recognised. What is typically neglected, in this recognition, is the simple fact of its non-singular semantic *basis*.

[6] In an earlier version, the table first appeared in Laycock, H. (1998). Words without Objects. *Principia* 2(2), 147–182.

3

Now just this negative account has been affirmed in recent work by Tom McKay. McKay (2008) notes that while NCNs are indeed on a par with plural nouns in respect of their non-singularity,

> plural discourse has natural semantic units that are the same as those of singular discourse, but stuff discourse has no natural semantic units, and reference and predication seem to proceed on a different model than that of an individual and a property.[7]

In consequence, in the case of words like "water", he urges that

> we should not expect a successful reduction to singular reference and singular predication, something that the application of traditional first-order logic would require. When we say that water surrounds our island, our discourse is not singular discourse (about an individual) and is not plural discourse (about some individuals); we have no single individual or any identified individuals that we refer to when we use "water". (McKay, 2008, p. 310)

We have, in short, no *individuals* when using "water"—and to this extent, at least, McKay and I are in complete agreement on semantics. The point is no mere point of grammar; as McKay puts it, "We seem to be in new territory ontologically, not just grammatically". But lacking individuals in this domain, how are we to *positively* think of "what there is"? McKay appears to have a clear and simple answer to the question—unsurprisingly, an answer which reflects the form that he believes the question needs to take; and the question to be asked in this connection is an entirely conventional one. The question, so he urges, is that of

> what the subject of the fundamental NCN predication is, i.e., of what satisfies "x is water", especially if we understand quantifier phrases on anything like a standard model. We must either understand what values x can take, or provide some other analysis of quantifier phrases.
> (McKay, 2008, p. 311)

---

[7]In his Critical Notice (McKay, 2008) of (Laycock, 2006).

For him, the crucial ontic question is just this: Of what concrete *subject-expressions* do we fundamentally assert or predicate "is water"? And the answer to this question seems as clear as clear can be—the substituends for "x", in expressions of the form "x is water", can only be expressions with the form of *this stuff*—*this water*, just for instance. And it is in just these terms that McKay answers his question regarding the subject of the fundamental NCN predication: we must, he observes, be "talking about *some stuff*".

Here, McKay's view coincides exactly with the object-grounded doctrine of concept-instantiation. "Some stuff" is rightly understood, by both, as designating some unspecified although determinate amount of stuff of one kind or another—something which may serve as basic subject. But the orthodoxy treats *some stuff* as semantically singular; and on this crucial point, McKay departs from orthodoxy. If the semantics of NCNs express no object-concepts, the conclusion has to be, as he remarks, that we are

> talking about some stuff, not a thing or some things, and in *that* way, mass reference and predication are ontologically more significant than plural reference and predication. (McKay, 2008, p. 311)

This then, in McKay's view, is the central logico-ontic difference between stuff and things. There has to be *some* kind of basic subject for a non-count predication; but lacking any range of distinct countables or objects, this basic subject can be nothing other than *some stuff*; and since this stuff is not itself a single object, entity or thing, it follows, as McKay bluntly puts it, that if "we are not to reduce mass talk to thing talk by introducing a realm of entities, it seems that we must somehow provide for talk about non-entities".

4

McKay speaks of "understanding what *values* x can take", for open sentences like "x is water". But now variables are no mere placeholders for linguistic substituends: as something *taking values* in the first place, variables are inseparable from the general ontic category to which those values necessarily *belong*—the category of objects, individuals or things. Precisely this is the core idea of Quine's maxim,

"To be is to be the value of a variable". In consequence, McKay's account is a problematic story to relate: recall Russell's key contrast of grammatical and ontic *subjects*. With a non-singular but essentially *plural* predication, the question "What is the subject of such a predication?" is readily answered in *grammatical* terms by pointing to the plural noun phrase. But if the question is asked about the *content* of this reference, the answer is that there simply isn't *one*. Grammatical subjects apart, an *ontic* subject has to be a *single* unit, whereas in plural predication, there are *several* subjects, not just one; and in a non-count predication, there are *neither* several subjects nor just one.

To speak *in the plural* of "non-entities" is to speak of a *plurality* or *multiplicity*, and so of a plurality or multiplicity of *units*—in other words, again, of *entities*—and therefore is intrinsically paradoxical. *Subjects* must at any rate be countable; *each* subject must be countable as *one*, and subjects must be *either* one or *many*. But "much"— the quantitative adjective for stuff—is *neither* one nor many. Against McKay's proposal, to say that there can be no ranges of non-entities is just an outright truism. There is absurdity in the notion of a domain of non-units, "each" of which is just some stuff. If *there is* something, of which there can be several distinct occurrences, then these "somethings" cannot fail to count as distinct entities. "The much" can therefore not be understood as a plurality of "subjects". If we agree that there can, as a matter of logic, be no such thing as the object of a non-count reference, no such single *object* as "the much", then by that same token, there can be no such thing as the *subject* of a non-count predication, no such single *subject* as "the much". If there can be no such things as instances, then there can be no such *subjects* as the designata of "some stuff".

Again, in understanding predications with the form of "x is water", McKay puts the question of "what *values* x can take". But if, as he himself urges, "we are talking about some stuff, not a thing or some things", then we can hardly continue to speak of distinct *values* of variables at all. The semantics of "value", as with those of "entity" or "subject", are themselves the semantics of a count noun. There can be no notion of *a value* which is not that of a single countable or unit, a single entity or thing. The concept of a value just *is* the concept of a single individual or thing, of whatever putative category or kind; and "values", truistically, is plural, as "a value" is semantically singular. But yet "some stuff", as McKay himself affirms, is semantically neither singular nor plural. Where predication involves an NCN, there just cannot be a question of what *values* x can take. If "we are in new territory ontologically", then we need a more coherent understanding of what its novel features are.

Rather than asking ontically question-begging questions such as "What satisfies 'x is water'?"—in which x is taken to be a variable carrying ontological commitment, and so demands no mere substituend, but an individual *value*—we need to pose the neutral question, of *what* the logical form of non-count assertions of existence might be; and whether assertions of existence must always have a "subject" in this sense. A formal system based on the idea of quantification answers this question in the affirmative. But it is not obvious that this answer is correct—and given that NCNs are neither singular nor plural, the answer *cannot* be correct. Since an amount of stuff is bound to be collapsed into a unit, if it is to be treated as an ontic subject in the first place, the conclusion seems unavoidable that there simply are no ontic subjects within this domain. An ontic subject can be nothing other than a single entity or thing—singular reference is the only form of reference whose *semantic form* comes with an *ontic subject* as its content. And, since NCNs are irreducibly non-singular, no referential non-count sentence has an ontic subject, or in general, any metaphysical significance. The semantic form of referential sentences *cannot express* in what the existence of a kind of stuff most basically consists. The issue inevitably draws in quantified sentences, since the form of such assertions includes placeholders for referential terms. Are there, then, sentences which are *neither* quantified nor of referential, subject / predicate form? Not only do such sentences exist; they are extremely commonplace.

Around half a century ago, Peter Strawson (1959) made some unusually interesting and challenging logico-metaphysical observations on what he called *feature-placing* sentences. The sentence "There is water here", so he remarked, is such as to "neither bring particulars into our discourse, nor presuppose other areas of discourse in which particulars are brought in" (Strawson, 1959, p. 203). Such sentences are marked by the syntactical—and also, as I'll argue, semantical—absence of quantifiers or determiners, and are thereby classified as *bare*.[8] Their widespread use and distinctive significance notwithstand-

---

[8]The terminology of "bare" nouns or noun phrases originates with Chomsky

ing, bare sentences in general—and non-count sentences in general, bare or not—are massively neglected; and more especially within philosophy than in linguistics. To understand these sentences, we must return first to the object-concept.

5

Now Table 1 is just one part of the story of the categories or concepts with which it is concerned. The table consists in a highly schematic taxonomy of *noun-types*; but concepts are distinct from words and sentences. Eluding direct inspection, they have what may be called semantical *embodiments* in sentences. One and the same concept is expressed in diverse forms of words in diverse languages; but also in a single language. "Happy", the adjective, and "happiness", the noun, express no distinct *concepts*, though the terms have semantically distinct functions. Even less do we suppose that "woman" and "women" express distinct concepts, since one is semantically singular, the other plural. Singular and plural forms of nouns express no differences of concept; the difference is essentially semantical, not conceptual. Just this point indeed was made regarding "object" in itself: whether we speak of *objects* or *an object*, one and the same underlying concept is expressed. One and the same category appears in box $\{4,1\}$ and in box $\{3,2\}$ of Table 1. This too is the force of Russell's remark—in response to Leibniz' maxim "Whatever is, is one"—that "Whatever are, are many". The category of objects in itself is *neither* singular nor plural, though it may be represented or expressed semantically in *either* form.

But further, while there is a *single* form of statement in the singular, plural statements take two logically distinct forms, in a contrast which is instructive for the case of NCNs. Consider, then, the contrasts of the existentially committed sentences below:

(a) A student is waiting to see you $\{4,1\}$

(b) Some students are waiting to see you $\{4,2\}$

(c) Students are waiting to see you $\{3,2\}$.

(1975).

	1. Singular ("one")	2. Non-singular ("not-one")
3. Bare, or non-identity / non-ref		"Objects" "Stuff"
4. Non-bare, or identity / ref	"An object"	"Some objects" "Some stuff"

Table 2

Notice that the form of (c) is that of a bare plural sentence—one which matches and illuminates a Strawsonian bare non-count sentence. And these semantic contrasts—along with parallels involving NCNs—may be represented in a tableau not unlike Table 1, where the aim is now to regiment the structural homologies of CNs and NCNs, and not to emphasize their differences, see Table 2.

The focal issue is the contrast of both (a) and (b) with (c). These sentences each have their quantified truth-conditional equivalents; however, there is also an important *non*-truth-conditional semantic difference between both (a) and (b), on the one hand, and (c) on the other. The difference distinguishes (a) and (b), but not (c), from the existentially quantified sentences of the predicate calculus (notwithstanding the fact that these are indeed existential, as opposed to referential, sentences). Along with having fully fledged referential counterparts, both (a) and (b) display what I'll call a certain anaphoric, "identity involving" or "cross-referential" force or use potential. For instance, if (a) is coupled with or followed by

(a′) She has been waiting to see you all morning,

then it's clear that the *same* individual is at issue. By the same token, (b) may be coupled with or followed by a plural counterpart of (a′). Yet, the quantified truth-conditional equivalents of these sentences—in the case of both (a) and (c),

At least one student is waiting to see you

—can have no identity-involving anaphoric link with other sentences like (a′); no explicitly quantified existential sentence with the form of "At least one ... ." can have such potential.

Furthermore, not only is this identity-involving potential or force something which the corresponding truth-conditionally equivalent formal sentences do not possess; it is a potential which cannot but be grounded in the *semantics* of (a) and (b). Both (a) and (b), as natural-language sentences, are linked to potential *subjects*, but (c)—*qua* plural, crucially—is not. It carries the least semantic baggage: it is numerically neutral, as (a) and (b) are not; and its semantic content is *identical* with that of its truth-conditional equivalent. In effect, in claiming only "at least one", (c), unlike (a) and (b), states *only* that the relevant concept is instantiated: no determinate number is attached to its instantiation. Hence with (c), potential subjects enter in only "indirectly" at this deeper, singular level; but as it stands, (c) is a *pure*, subjectless embodiment of the object-concept.

6

Compare this situation with the non-count case. The non-count boxes, lacking singularity, are only two, and as examples I suggest the following

(d) Some water is on sale at the well $\{4, 2\}$

and

(e) Water is on sale at the well $\{3, 2\}$.

In relation to anaphora and identity-potential, these match the corresponding CN sentences. Evidently, (d) calls forth *bona fide* anaphora, whereas (e) does not. For (d) there is

(d') It's been on sale all week (nobody wants it)

and for (e) there is

(e') It's been on sale all week (everyone's buying it).

The relationships of (d) and (e) are parallel to those of (b) and (c): (d) presupposes (e), but not vice-versa. Thus (d) implies determinate but unspecified amount—there may be just a liter left—whereas (e) is open-ended and wholly neutral, indeterminate in quantity. And like (e), a Strawsonian feature-placing statements such as

(f) There is water at the well

is both existential and unquantified. Sentences having this form, I suggest, constitute *pure* embodiments of the matter-concept—here the category of stuff itself appears directly, unmarked and unmodified by the presupposition of quantitative elements, identity or potential reference. And in this light, Strawson's claims about his feature-placing sentences do not appear outrageous. In contrast, the d-type sentence has identity and quantity tacked on. Sentences (b) and (d) both involve a general concept *along with* either number or amount. "Some" marks an element of empirical information, adventitious from the standpoint of the ontic categories themselves; these may be called "impure" or *hybrid* semantic / ontic forms which lay the semantic basis for non-singular identity-statements. In contrast, the form of (e) represents the pure or subjectless embodiment of the matter-concept.

Now the (c)-type bare *count* sentences—

(g) There are students waiting to see you

—I've proposed, also lack subjects and identity, as is evident from the non-identity-involving, pseudo-anaphoric

(h) They've been waiting to see you all week.

But there is of course a crucial difference between these and the bare non-count sentences; for here, the bare unquantified non-singular form, essential to its non-referential (non-identity involving / no subject) status, *reduces* to a quantified singular form, and open thereby to potential subjects. But this *cannot*, plainly, happen with the non-count form.[9]

7

Both the idea of an amount of matter, and the idea of a number of objects, involve combining the ontic categories at issue—those of *objects* and of *matter*—with a notion of specific *quantity*, a notion of how many or how much. The idea of *a number of* objects self-evidently

---

[9]The two categories may be drawn more closely together through the use of collective count noun predications, as with "Students are milling around outside the building"; but the basic point remains the same.

combines the neutral idea of objects *simpliciter* with the further idea of determinate but unspecified *multiplicity* or number. Similarly, the idea of an amount of stuff combines the neutral idea of stuff or matter *simpliciter* with the idea of determinate but unspecified amount. And corresponding to this difference is a difference in the relevant idea of quantity. With objects and count nouns, we understand how to "construct" the idea of a number of objects from that of single objects; it is essentially a matter of addition; there is this one and that one and that one. But this is plainly impossible with NCNs. The question therefore arises of just how the *transition* is made, from the idea of stuff to the idea of an amount of stuff—how is this idea generated? Perhaps the most obvious suggestion is that it must "come in from outside": the paradigm case might involve a container. The physical idea of a *closed system* seems essential to any guarantee of identity, since identity is ungrounded in the concept in itself.

The idea may be approached, via the question of when and how an amount of matter might come into being or cease to be. The idea is not commonly addressed; but among those who have at least touched upon the issue, there is a tendency to treat it as *eternal*, or quasi-eternal. Richard Sharvy (1983) remarks, somewhat uncertainly, that

> The matter of my car cannot lose any part, but my car can. Whatever matter is, and whether or not any of it every comes to be or passes away, no bit of it ever changes any part, so long as that bit exists. Probably, there is no coming to be and passing away of matter, but only mingling and separation... Where is the snow of yesteryear? All around us and in us. (Sharvy, 1983, p. 237)

Matter, so the thought goes, cannot be simply "gotten rid of" via any sort of disintegration or dispersal. It is, after all, what substances disintegrate *into* when they cease to be; where then could the stuff itself be thought to go? The ancient principle, *ex nihil, nihilo fit*, is in effect the other side of just this coin.[10] (And yet, with objects,

---

[10] It looks *prima facie* as if such "things" would *have* to be eternal. If "they" were not, then either they would have to be capable of absolute annihilation without remainder, or else there would have to be *another* category of being into which they were "resolved"; but what could that conceivably be? The answer "energy", for instance, seems to invite the reply both that energy is just another kind of

identity and persistence depend crucially on form and *not* matter!)

To further pursue the question, I want to contrast two kinds of mixture—mixtures of things , and mixtures of stuff. And first, where things of distinct *types* are mixed—nuts are mixed with raisins, say—the challenge of tracking and re-identifying individual objects in the resulting mixture need not be addressed. If we wish to separate the raisins from the nuts, it's enough to separate the objects of one kind from the objects of the other kind. But with objects of the same kind we must track every individual from at least one of the groups to be mixed. Likewise, where different kinds of *matter* are involved, the possibility cannot be ruled out that the stuff of each distinct kind may be separated out or recovered from the mixture by some special process. When however aggregates of stuff of the *same* kind are mixed, then if the stuff in question is fluid, lending itself to thorough mixing, such a possibility will not exist. Situations of both kinds are considered in what follows.

Experiment 1 We begin with a jug full of peanuts, and two bags labelled "P" and "R". Pour the peanuts from the jug into the two bags. Now add the contents of these two bags into a suitably sized bowl and stir well. The nuts from the two bags are then well mixed, such that close to each nut from bag P is at least one nut from bag R, and vice-versa. None of the individual nuts, we may stipulate, is destroyed by the mixing process. In this case, we have provided ourselves with a a guarantee that the nuts which were in P do continue to exist once in the bowl, but there is a good sense in which the nuts which were in P are not capable in principle of being recovered, identified or distinguished, once they are in the bowl.

While there would be no need to track each of the individual nuts and raisins in a mixture of nuts and raisins, in order to be sure of separating out the contents into a bag of nuts and a bag of raisins once again, there is no such easy solution to the challenge in this case. Since the objects are all of the same kind, we must rely purely and simply on the distinctness and re-identifiability of each and every individual object. There are *these nuts* here, in P, and *those nuts* there, in R; but while these nuts are currently *collectively* distinct

---

stuff, if not "material" in the strongest, space-occupying sense—and that some conservation principle would still need to apply.

from those, they are not intrinsically distinct from those—other than individually. And if we may suppose that the individual nuts are, for all intents and purposes, *indistinguishable* from one another, then we have no option but to rely on the tracking of each of the individual nuts from either P or R, through the entire process of mixing. The point reflects the fact that a principle of distinctness is a built-in, essential or intrinsic feature of each and every *individual* concrete object, but it is no more than an accidental or extrinsic feature of a number of objects collectively. An individual object cannot lose its individual distinctness without ceasing to be; but a number of objects can lose their collective distinctness without ceasing to be. The collective distinctness of a number of objects is a function of their *constituting* a discrete aggregate or collection.

Experiment 2 Take a jug full of pure water, and two glasses labelled "S" and "T". Pour a small amount of the water into each of S and T, so that each glass is less than half full. Then take glass S, and pour the water from that glass into T, so that glass T now contains the water from each—and equally, from both—of the two glasses. Stir well.

Now if there is any reason to suppose that the water from each glass persists in T in any more robust sense than this—namely, that it was poured *into* it—those reasons are not immediately obvious. Consider first the contrast of the two experiments. In both cases, we can distinguish between the contents of two separate regions of space—the nuts here and the nuts there, the water here and the water there. And in neither case does the mixing or mingling involve the mixing or mingling of distinct kinds. Collectively, in experiment one, the nuts here are distinct from the nuts there; and analogously for experiment two, the water here is distinct from the water there. Just so long as they remain in P, the nuts in P are distinct from the nuts in R not merely individually, but also collectively. But there is plainly no intrinsic *collective* contrast or distinction between the nuts which happen to be in the one place and the nuts which happen to be in the other place. That is precisely why they can be mixed or mingled. To belabour the obvious, the only ultimately relevant distinction in this case is that between individual nuts.

Experiment two, on the other hand, presents a scenario involving non-atomic words for stuff. Very roughly, the contrast is that between words like "wine" and "water", and words like "furniture" and "footwear".[11] Here, *ex hypothesi*, there are no such semantically determined individual units to rely on. In aggregate and as such, the water *in* T is, of course, distinct from the water *in* S—there are, in a non-technical sense of "mass", two spatially distinct *masses* here, two distinct bodies of water, the boundaries of each being constituted by the interface of the water and the glass. But there is nothing corresponding to the intrinsically distinct "atoms" or individual nuts, and once these distinct masses merge, such that their externally imposed distinctness disappears, there is no further basis for distinctness whatsoever. The water in each of the glasses is distinguished by its boundaries, which are essentially those of each of the two distinct *glasses*—and which thereby *constitute* the water in each glass, *as such*, as distinct. In the absence of any reason to embrace an a priori atomism for the case of non-atomic NCNs, however, there can be no a priori basis for a belief in some other, *intrinsic* principle of distinctness and identity.

The concepts here at issue are entirely consistent with—which is to say, they permit—the possibility of total fusion. There is nothing in these concepts to preclude the possibility of such a fusion. Insofar as the water from glass S was added to glass T, there is undeniably some sense in which the water from each of the glasses can be said to be now *in* glass T. But this justifies no belief in either the continued or potentially distinct existence of the water which was in either S or T. There is no basis in the scenario for a belief that the water which was in either glass must, *a priori*, be capable of being recovered (or could even re-appear by some very lucky accident) if the water is again poured into distinct glasses. The identity in question cannot but be adventitious, "external" as against intrinsic. The thought that the water which was in S is now in T is a possible thought, *only* because of what had happened in the past; it need correspond to no *actual* distinction in the contents of glass T. As Locke famously insists, a principle of physical distinctness or mutual exclusiveness is a built-in feature of the concept of a concrete body or material substance. But in the case of a number of bodies collectively, or of an amount of stuff in aggregate, the distinctness is no more than an accidental or adventitious feature; the possibility of mixing or mingling is inherent

---

[11] For further discussion of this issue, see (Laycock, 2006, ch. 1).

in these cases.

In other words, the supposed identity of the water now in T with the water previously in T and S is not a function of the persistence of the water previously in T and the water previously in S. That supposed identity is not a function of the persistence of *any* "sub-amounts" of water. It is, plainly, a necessary condition of the identity of the water now in T with the water previously in T and S that the *amount* of water now in T is identical with the sum of the *amounts* of water previously in each of T and S. But the amount of water in T, like the number of people in this room, is a kind of abstract magnitude, and not a concrete "portion", "quantity" or "mass". A principle for the conservation of the amount of water involved in such a fusion is not a principle for the conservation or persistence of some concrete stuff itself.

Given a stipulation that the context of the experiment is that of a closed system, whereby nothing leaves or enters the domain of the experiment, it is reasonable to believe that the amount of water before and after mixing is the same. The volume (or rather, the somewhat elusive *measure-independent* amount, assuming that such a thing exists) of water in glasses S and T before the "merger" cannot fail to be the same as that in glass T past the merger; but that is a very different matter. Appropriately qualified, that would certainly appear to be a kind of a priori truth.

Putting to one side the atomic NCNs, there can be no more to our conception of concrete quantitative identity, as applied to stuff of any particular kind, than that of spatio-temporal distinctness—whereby, in the nature of the case, this distinctness needs to be externally imposed or generated, and is not intrinsic to the stuff. When addressing mixture in the context of a single kind of stuff, the idea that (what happens at some particular time to be) a concrete, isolated amount of matter must continue to exist no matter what, seems to be an illusion. Not only are there no such *things* as amounts of matter; but there is no reason to believe that anything which could be truly said to be some stuff or some matter of one sort or another should have any built-in tendency to perpetuate itself. The nature of a material "conservation principle" regarding the overall quantity or amount of matter involved in any change (a principle akin to that of the quantitative conservation of mass/energy) obscure as may seem to be, requires no such persistence.

8

Now a river is not a glass of water; and with Russell's notion of denoting in mind, gesturing towards the water in the Neva—"the" water in the Neva—is not pointing out a re-identifiable, discrete amount of water.[12] Nevertheless, we may imagine applying a system of spatio-temporal co-ordinates to divide up the river into discrete volumes of space, however large or small, enabling us thereby to *denote* the water in any precise region r at any moment of time t. And this will, we may suppose, be some particular *amount* of water. There is however no reason to suppose that the water thus denoted must persist, whether in the same or any other regions, at any other time t'. The mingling envisaged in the case of glasses S and T is the norm rather than the exception for the water in a river. We might perhaps imagine discrete volumes of water, of enormously diverse shapes and sizes, temporary masses, generating and degenerating as they move, and resulting in the flow of water from the hills and mountains to the sea. And so, indeed, on a global scale, for water in general and as such. Insofar as water persists, this cannot, except *per accidens*, consist in the persistence of discrete amounts. (Indeed, for the logico-semantic reasons earlier rehearsed, there can be no such multiplicity of *things*; adopting the *facon de parler* in this manner serves just to reinforce the formal point). The persistence of water is not the persistence of discrete amounts; such as it is, it is the persistence of water *simpliciter*.

If indeed a solution it be, this explanation of the absence of ontic subjects, for the categories of non-count nouns and stuff, will surely seem unsatisfying. But it will seem so precisely because it says so very little. But again, if I am right, it says so very little for there is no more to be said. This, it seems to me, and nothing more, is how things are.

References

Chomsky, N. (1975). Questions of form and interpretation. *Linguistic Analysis*, *1*(1), 75–109.

---

[12] Like any human activity, pointing or gesturing takes time; and the water in the river provides no stable target for full-fledged identifying reference; hence the usefulness of denoting.

Hacker, P. (1979). Substance: the constitution of reality. *Midwest Studies in Philosophy*, *4*, 239–261.
Laycock, H. (2006). *Word without objects*. Oxford: Clarendon Press.
McKay, T. (2008). "Critical notice" of Words without objects. *Canadian Journal of Philosophy*, *38*(2), 301–323.
Russell, B. (1903). *The principles of mathematics*. Cambridge: Cambridge University Press. (Reprinted, London: Allen & Unwin, 1937)
Sharvy, R. (1983). Mixtures. *Philosophy and Phenomenological Research*, *44*(2), 227–237.
Strawson, P. (1959). *Individuals*. London: Methuen.

Henry Laycock
Department of Philosophy, Queen's University,
Kingston, Ontario, Canada K7L 3N6
e-mail: laycockh@queensu.ca
URL: http://post.queensu.ca/~laycockh/

Burge's Contextual Theory of Truth and the Super-Liar Paradox

Matt Leonard*

Abstract

One recently proposed solution to the Liar paradox is the contextual theory of truth. Tyler Burge (1979) argues that truth is an indexical notion and that the extension of the truth predicate shifts during Liar reasoning. A Liar sentence might be true in one context and false in another. To many, contextualism seems to capture our pre-theoretic intuitions about the semantic paradoxes; this is especially due to its reliance on the so-called Revenge phenomenon. I, however, show that Super-Liar sentences (where a Super-Liar sentence is a sentence which says of itself that it is not true in any context) generate a significant problem for Burge's contextual theory of truth.

Keywords: liar paradox, contextualism, super liar, revenge, truth

1 Introduction

The first sentence of this paper is false. Why is the first sentence such a problem? Well, suppose that the first sentence is indeed false. If we judge the first sentence false, then it seems to be true, because it (truly) says of itself that it is false! Now, suppose that the first sentence is true. If the first sentence is true, then what it claims is

---

*The author wishes to thank Michael Glanzberg, Adam Sennet, Terence Parsons, Mark Balaguer, David Pitt, Robbie Hirsch, Jacob Caton, Reuben Stern, Goncalo Santos, Ray Jennings, Tyrus Fisher, Michael Hatcher, Michael Anderson, Zeph Scotti and audiences at UC Berkeley, Georgia State University, Arizona State University, and the 2011 Logica Conference for helpful discussions and comments.

true. But it claims that it is false; so it must be false! The problem, then, is that we have a sentence, which seems to be both true and false. But (just about) all of us believe that contradictions cannot be true!

Given the following two rules of inference,

Semantic Ascent: $\alpha \vdash \text{Tr}(\ulcorner \alpha \urcorner)$

Semantic Descent: $\text{Tr}(\ulcorner \alpha \urcorner) \vdash \alpha$[1]

and the following instance of the Liar,

β $\ulcorner \beta$ is not true.$\urcorner$

the Liar's formal proof sometimes runs the following course:

(1) $\beta = \ulcorner \beta$ is not true.$\urcorner$ [Given]
(2) Assume $\ulcorner \beta$ is true.$\urcorner$ [For *Reductio*]
(3) $\ulcorner \beta$ is not true$\urcorner$ is true.   [Substitutivity, (1) and (2)]
(4) $\ulcorner \beta$ is not true.$\urcorner$ [Semantic Descent from (3)]
(5) $\ulcorner \beta$ is not true.$\urcorner$ [*Reductio* (2)–(4)]
(6) Assume $\ulcorner \beta$ is not true.$\urcorner$ [For *Reductio*]
(7) $\ulcorner \beta$ is not true$\urcorner$ is true.   [Semantic Ascent from (6)]
(8) $\ulcorner \beta$ is true.$\urcorner$ [Substitutivity, (1) and (7)]
(9) $\ulcorner \beta$ is true.$\urcorner$ [*Reductio* (6)–(8)]
(10) $\ulcorner \beta$ is true$\urcorner$ and $\ulcorner \beta$ is not true$\urcorner$.   [(5) and (9)]

One relatively recent response to the Liar is provided by those who endorse a contextual approach to the semantic paradoxes. In general, contextualism is the view that there is an indexical element involved in the reasoning process of the Liar paradox; given a token of the Liar sentence, the extension of 'true' is contingent upon the context of utterance, and in some theories, the intentions of the speaker. Truth is an indexical notion. If a Liar sentence is not true in some context $\Gamma 1$, then the same Liar sentence will be true in a context $\Gamma 2$, where

---

[1]Semantic Ascent should be read as "From α, you may validly infer that α is true." Semantic Descent should be read as "If α is true, then you may validly infer α."

Burge's Contextual Theory of Truth 143

$\Gamma 2 > \Gamma 1$. There are many different contextualist theories of truth; I will, however, be looking at only one. In section 3, I will explicate Burge's (1979) theory. He claims that the extension of the truth predicate varies with shifts in context. What I hope to do in this paper is present a modest and novel worry for Burge's theory of truth. In section 4, I will mention some worries I have with his theory (and in particular, how his theory deals with the so-called Super-Liar paradox). I'll end by trying to show that the theory as it stands falls into somewhat of a dilemma. But first, let me provide some context by looking at Tarski, Kripke, and the so-called Revenge phenomenon.

2 Tarski's hierarchical theory and Kripke's paracomplete theory

Tarski maintained that the threat of paradox emerges when the truth predicate for a language L_1 resides in L_1 itself. This is why natural language generates paradox. Thus, he proposed that the truth predicate for a language L_1 must be placed in a metalanguage L_2. If we start with an interpreted language L_1, which excludes a truth predicate, we can then add a truth predicate to form L_2 and make claims regarding the veracity of sentences in L_1. For instance, for a sentence $\ulcorner \phi \urcorner$ in L_1, we can claim in L_2 that $\ulcorner \phi$ is true.$\urcorner$ The hierarchy is infinite. For any sentence $\ulcorner \phi \urcorner$ and any level n, we can only claim that $\ulcorner \phi$ is true$\urcorner$ in L_{n+1}. How does this solve the Liar paradox? It solves the Liar paradox because it blocks the formulation of a Liar sentence. Since there are no Liar sentences, there is no paradox.

Though Tarski (1933) did successfully block the Liar in giving his definition of truth, most people (I think rightly) want to say that while Tarski's definitions of truth and denotation are fruitful for metalogic, they are too restrictive for our ordinary notions of truth and meaning. One's theory of the semantic paradoxes should match our pre-theoretic intuitions about natural language, rather than block paradoxical sentences in an artificial language. Hence, Tarski's solution to the Liar is too restrictive.

A more recent (and popular) theory is the paracomplete solution to the semantic paradoxes. Paracomplete solutions maintain that Liar sentences do not have a truth-value (they lack truth conditions). Perhaps the most influential response endorsing truth-value gaps is

Kripke's (1975) theory of truth. Kripke begins with a classical language that lacks a truth predicate.

We should think of an interpreted language L as an ordered triple $\langle \mathcal{L}, \mathcal{M}, \sigma \rangle$, where $\mathcal{L}$ is the syntax, $\mathcal{M}$ is a model that provides an interpretation to the nonlogical vocabulary, and σ is a valuation scheme. Classical languages are characterized as having the following set $\mathcal{V}$ as their 'semantic values': $\{1, 0\}$.[2] Let L_0 be a classical language. In L_0, $\mathcal{M} = \langle \mathcal{D}, \mathcal{I} \rangle$, where $\mathcal{D}$ is a non-empty domain and $\mathcal{I}$ is an 'interpretation-function' which assigns to each name of L_0 an object from $\mathcal{D}$ and assigns to each n-ary predicate an element of $\mathcal{D}^n \to \mathcal{V}$, in other words, a function taking n-tuples of $\mathcal{D}$ and yielding a truth value, i.e., a semantic value 1 or 0.[3] The extension of an n-ary predicate F contains all n-tuples $\langle a_1, \ldots, a_n \rangle$ of $\mathcal{D}$ such that $\mathcal{I}(F)(\langle a_1, \ldots, a_n \rangle) = 1$, or colloquially, the set of things of which F is true. The valuation scheme for classical languages is τ (dubbed τ for Tarski), where a disjunction is true iff one of its disjuncts is true, a conjunction is true iff both of the conjuncts are true, and so on.

Kripke constructs a non-classical language using Strong Kleene logic in the following way.[4] $\mathcal{V}$ is expanded to $\{1, \frac{1}{2}, 0\}$ and so our new language L_1, $\langle \mathcal{L}, \mathcal{M}, \kappa \rangle$, where κ is the new valuation scheme, is now a three-valued non-classical language. For L_1, model $\mathcal{M} = \{\mathcal{D}, \mathcal{I}\}$, where $\mathcal{I}$ does the very same thing in L_1 as it did in L_0, except now it assigns to n-ary predicates elements of $\mathcal{D}^n \to \{1, \frac{1}{2}, 0\}$. We want to conceive of predicates in terms of extensions and antiextensions. As in the classical language, the extension of an n-ary predicate F contains all n-tuples $\langle a_1, \ldots, a_n \rangle$ of $\mathcal{D}$ such that $\mathcal{I}(F)(\langle a_1, \ldots, a_n \rangle) = 1$, or colloquially, the set of things of which F is true. The antiextension of an n-ary predicate F contains all n-tuples $\{a_1, \ldots, a_n\}$ of $\mathcal{D}$ such that $\mathcal{I}(F)(\langle a_1, \ldots, a_n \rangle) = 0$, or colloquially, the set of things of which F is false. L_1 leaves open the possibility of some n-tuples not falling in either the extension or antiextension of F; in this case, we say that F is *undefined* for some n-tuple. Let F^+ and F^- be the extension and antiextension of F. L_0 and L_1 both agree that nothing exists in both the extension and antiextension; or, $F^+ \cap F^- = \emptyset$. They differ in the

---

[2] Let '1' represent 'is determinately true', '0' represent 'is determinately false', and when I mention it shortly, let '$\frac{1}{2}$' represent 'is undefined'.

[3] And likewise assigns to each n-ary function-symbol an element of $\mathcal{D}^n \to \mathcal{D}$, i.e., an n-ary function from $\mathcal{D}^n$ to $\mathcal{D}$.

[4] Here I'll rely on a nice summary of Kripke in (Beall, 2007).

following way. As opposed to L_0, L_1 holds that there can be an x such that x is undefined for F; L_1 denies that, necessarily, $F^+ \cup F^- = \mathcal{D}$. In other words, L_1 denies that every sentence is in the extension or antiextension of the truth predicate.

Kripke then constructs his so-called fixed-point language. Kripke begins with (the classical) L_0, which lacks a truth predicate and extends it to (the non-classical) L_1, which contains a truth predicate. Unlike Tarski's theory, the truth predicate can be applied to every sentence of L_1 (including all of the sentences of L_0). In the above paragraph, I mentioned that L_0 is to be interpreted with a classical model $\mathcal{M}_0$. Kripke proposes to build up a model $\mathcal{M}_1$ for the expanded L_1. Kripke employs an inductive method here. Start with an empty extension and an empty antiextension. Start throwing in true sentences to the extension and false sentences into the antiextension. Eventually, Kripke shows, we will arrive at a level (this is going to be a transfinite level) where adding any more sentences to the extension and antiextension will cease to be 'productive,' i.e., it reaches a *least fixed point*. Liar sentences do not appear in either set, and thus are viewed as 'gappy.' What Kripke seems to have shown is that (i) a language can contain its own truth predicate and (ii) Liar sentences come out lacking a truth-value.

This common sort of response to the Liar, however, has been met with a serious problem. It is often referred to as the 'Revenge of the Liar,' or 'Strengthened Liar reasoning.' The revenge problem is not really a new problem; it is simply another instance of the Liar masked for truth-value gap responses to the original Liar. Technically, the original Liar is of the following form,

β_{OL} β_{OL} is false.

When met with the original Liar, one can just claim that β_{OL} cannot be true and it cannot be false; 'No problem, β_{OL} lacks a truth value'. However, consider again an instance of the Strengthened Liar,

β_{SL} β_{SL} is not true.

The Strengthened Liar is supposed to show that β_{SL} cannot be true, cannot not be true, and cannot not have a truth-value. So what is the revenge problem? If β (from now on, just take β to have the strengthened form) is neither true nor not true, then in particular it is not true. But if it is not true, then it seems that β is not true (since that is what β seems to tell us). Therefore, β seems to be true in an important sense; β is true "after all"! The Strengthened Liar presents the problem in a more intuitive way than the original Liar. So there seems to be a reformulated paradox for 'gappy' theories.

3 Burge and contextualism

Burge wants to distance himself from both the Tarskian and Kripkean solutions to the Liar. He rejects the former for the same reasons that many people do, as I've mentioned above (i.e., it is too restrictive for our ordinary notion of truth, and so on). He rejects the latter, i.e., truth-value gap theories, because of the revenge problem. As a result, he posits a hierarchical theory that, though similar in some respects to Tarski's, differs by attempting to meet some of the pre-theoretic semantic intuitions Tarski's theory did not account for. In particular, he does this by claiming that the truth predicate is indexical and that its extension shifts from context to context. Now, in "The Concept of Truth in Formalized Languages," Tarski sought to block the Liar by assuming that he was dealing with some purely extensional concept of truth, not our ordinary notion of truth. In fact, he argued that natural language was inconsistent and inevitably generated the Liar. Burge, on the other hand, is interested in our ordinary/natural notion of truth. He wants to give a theory concerning our ordinary notion of truth which can block the Liar paradox *as it occurs in natural language*. He frowns on theories which simply block liar sentences in artificial languages with fancy technical ingenuities.

Burge argues that there is a hidden conversational implicature and a shift in extension (parallel with a shift of context) that occurs in Strengthened Liar reasoning. According to Burge, Strengthened Liar reasoning runs the following course:

Step 1:   An occurrence of a Liar like sentence.

Step 2:   The Liar sentence is not true.

Step 3:   The Liar sentence is true after all.

Most solutions to the Liar have either ignored such reasoning or attempted to block it by formal means. Burge, on the other hand, thinks a more satisfying approach is to interpret the reasoning so as to justify it. He thus takes the Strengthened Liar as a model for how we *should* think when confronted with the semantic paradoxes.

Consider the following very plausible scenario (and notice the corresponding Steps 1–3):

> Suppose I see a fake university professor enter a room and begin writing falsehoods on the blackboard. Suppose also that I think that I am in Room 398 and that the fake professor, at this moment, is in 399. So I write on the board at 11:30 A.M. on 6/24/11, (Step 1) "There is no sentence written on the board in Room 399 at 11:30 A.M. on 6/24/11 which is true as stan- dardly construed." However, unbeknownst to me, I am in fact the one in Room 399, and this is the only sentence written on the board. The usual Kripkean (or gappy) reasoning shows that this cannot have truth conditions; thus, it is not true. (Step 2) So there is no sentence written on the board in Room 399 at 11:30 A.M. on 6/24/1 which is true as standardly construed. But we have just stated the sentence in question. (Step 3) Thus, it is true after all.

The truth predicate used in this scenario is not some technical notion of truth (like, say, Tarski-truth). It is our ordinary notion of truth. This is the sort of paradox to which Burge is interested in providing a solution.

Burge wants to stipulate a formal system that defines a *pathological* sentence, as interpreted in a context. Burge stipulates that *pathologicality* is a disposition to produce disease for certain semantical evaluations. Thus, the Liar comes out pathological.[5] *Rootedness* is defined as the lack of pathologicality, i.e., a formula's being rooted means that it is nonpathological, and (roughly) that it has a truth-value.[6]

Burge then distinguishes extensions of 'true' by marking occurrences of them with subscripts beginning with i. In the Strengthened Liar case, for Step 2 Burge claims that the Liar sentence is not true. He marks this initial context of utterance 'true$_i$'. In Step 3, and from a broader application of truth, he claims that the Liar is true.

---

[5]The Truth-Teller (a sentence which says of itself that it is true) will also come out pathological.

[6]Rootedness is essentially the same notion as groundedness in Kripke's theory.

Burge dubs this context of utterance 'true$_k$,' where $k > i$. He argues that though pathological$_i$ sentences are not true$_i$, pathological$_i$ sentences are nonpathological$_k$, and thus true$_k$. All rootless$_i$ sentences are not true$_i$. So a sentence and its negation may both be not true$_i$, though one or the other will be true$_k$. Burge doesn't offer the following restricted Tarskian truth schema anywhere, but I presume this is a T-schema he would accept (which I'll subscript 'B' for Burge):

(T_B): ($\forall i$) If a sentence $\ulcorner \phi \urcorner$ is rooted$_i$, then $\ulcorner \phi \urcorner$ is true$_i$ iff p.

where $\ulcorner \phi \urcorner$ names any well-formed sentence in Burge's system and $\ulcorner p \urcorner$ is the sentence itself. Notice how Burge's theory interprets the Strengthened Liar:

Step 1: β $\ulcorner \beta$ is not true.$\urcorner$ [i.e., a Liar token.]

Step 2:   The Liar is not true.   [i.e., $\ulcorner \beta \urcorner$ is not true$_i$.]

Step 3:   The Liar is true after all.   [i.e., Step 2 is true$_k$.]

4   A dilemma for Burge's contextual theory of truth

The initial appeal of contextual approaches to the semantic paradoxes is that they accord with some of our intuitions about truth, and in particular, how to interpret the Strengthened Liar dialectic. As enticing as this appeal might be, there is a worry with Burge's contextual theory that throws doubt on whether this type of response to the Liar is, in fact, the right type of response. I want to mention both a general worry for all contextualist solutions to the Liar, and a specific worry with Burge's theory. The specific worry is just a problem with Burge's response to the general worry; so first, let me mention the general worry.

The general threat to contextualism emerges when the Strengthened Liar is reformulated in a way that explicitly refers to hierarchical contexts; this formulation is sometimes referred to as the Super-Liar. What type of response can the contextualist provide for sentences like 'This sentence is not true at any level, or in any context,' or sentences like $\ulcorner \psi \urcorner$?

ψ ($\forall i$) $\ulcorner \psi$ is not true$_i$.$\urcorner$

It seems that contextualism faces the same sort of paradox Tarski and Kripke face. Either $\ulcorner\psi\urcorner$ is not true$_i$ at any level i or $\ulcorner\psi\urcorner$ is true at some level n. Suppose it is not true$_i$ at any level. But that is just what $\ulcorner\psi\urcorner$ says of itself. Hence, $\ulcorner\psi\urcorner$ is true$_k$, where $k > i$ (i.e., $\ulcorner\psi\urcorner$ is true 'after all'). On the other hand, suppose $\ulcorner\psi\urcorner$ is true at some level n. If that is the case, then $\ulcorner\psi\urcorner$ should come out false at n, because it says of itself that it is not true at any level. In both cases, contextualism seems to be unable to account for $\ulcorner\psi\urcorner$.

Let me show even more explicitly the problem with $\ulcorner\psi\urcorner$ (using Burge's notation). Here I'll universally quantify over the extensions that can be applied to the truth predicate:

(1) $\ulcorner\psi\urcorner : \ulcorner(\forall i)\ulcorner\psi\urcorner$ is not true$_i\urcorner$ [Given]

(2) Assume $(\forall i)\text{Tr}_i\ulcorner\psi\urcorner$ [Reductio]

(3) $\text{Tr}_i[(\forall i)\neg\text{Tr}_i\ulcorner\psi\urcorner]$ [Substitution (2)]

(4) $(\forall i)\neg\text{Tr}_i\ulcorner\psi\urcorner$ [Semantic Descent (3)]

(5) $(\forall i)\neg\text{Tr}_i\ulcorner\psi\urcorner$ [Reductio (2)–(4)]

(6) Assume $(\forall i)\neg\text{Tr}_i\ulcorner\psi\urcorner$ [Reductio]

(7) $(\forall i)[\text{Tr}_{i+1}(\neg\text{Tr}_i\ulcorner\psi\urcorner)]$ [$\ulcorner\psi\urcorner$, Burge's Theory]

(8) $(\forall i)\text{Tr}_{i+1}\ulcorner\psi\urcorner$ [Substitution (7)]

(9) $(\forall i)\text{Tr}_i\ulcorner\psi\urcorner$ [Reductio (6)–(8)]

(10) $(\forall i)[\neg\text{Tr}_i\ulcorner\psi\urcorner \wedge \text{Tr}_i\ulcorner\psi\urcorner]$ [(5), (9)]

Contextualism seemed most plausible when it was allegedly able to circumvent the revenge problem; Super-Liars, at least prima facie, seem to immediately force contextualist truth theories back into paradox. Some philosophers think that this is a knock-down argument against contextualism. I do not intend to settle this difficult question, in this paper. However, I should note that Burge already knows that this version of the Liar can be generated against contextualism; in fact, he provides a response in advance in his original paper. What is puzzling is that no one seems to address his response. In what follows, I'll argue that the most devastating problem emerges when we put pressure on Burge's response.

Foreseeing the potential problem, Burge writes,

> Attempts to produce a 'Super Liar' parasitic on our symbolism tend to betray a misunderstanding of the point of our account. For example, one might suggest a sentence like (a), '(a) is not true at any level'. But this is not an English reading of any sentence in our formalization. Our theory is a theory of 'true', not 'true at a level'.
> (Burge, 1979, p. 192)

Burge wants to allow the schematic variables on the truth predicate to be contextually determined (by some Gricean process). But he doesn't want to allow quantification on them, in something like the way that type-theoretic levels, in type theories, do not allow quantification. They do not mark a quantifiable argument place on the truth predicate. It's not as if there is *really* a parameter there. The 'parameter' is really just being used as a label to indicate that there is some Gricean process going on. So if you think of it this way (where the truth predicate is immune from quantification) then you can't really formulate the Super-Liar because you can't formally quantify over contexts (or, extensions which are generated from contexts).

Recall that Burge is interested in giving a theory of our ordinary notion of truth. Suppose Ralph is walking down the street and shouts the following to a crowd of people,"To the sentence immediately following this one, I stipulate the name ⌜ϕ⌝. ⌜ϕ⌝ is not Tarski-true in L, in any of the transfinite metalanguages of L." The crowd would completely ignore Ralph. Why? Because his utterance is not ordinary English. Super-Liars, you might think, utilize some technical notion of truth and hence are immune from the relevant considerations.

I don't think that this is the case, however. I'll briefly argue that it is not inconceivable to construe a Super-Liar in ordinary language (using an ordinary truth predicate). Suppose we have a situation similar to the one I described earlier (with the professor who was a fraud). Suppose again that I walk into Room 399 but think that I am in 398 (and I also think that the fake professor is in 399). Suppose I write on the board at 11:30 A.M. on 6/24/11 one the following:

1. "There is no sentence written on the board in Room 399 at 11:30 A.M. on 6/24/11 which will ever be true."

2. "The sentence written on the board in Room 399 at 11:30 A.M. on 6/24/11 is not true in any context."

Burge's Contextual Theory of Truth 151

3. "The sentence written on the board in Room 399 at 11:30 A.M. on 6/24/11 is not true, no matter how you judge it (or, no matter how you look at it)."

I am not claiming that speakers understand the technical notion of contexts. All the speaker must do is utter one of the sentences above; and when she does, it seems obvious that we have some sort of natural language Super-Liar. But this is a problem for Burge. These above sentences seem to obviously include our ordinary truth predicate (in the same way that the example Burge provided regarding the fake professor includes our ordinary truth predicate). Burge's theory turns out inconsistent for such sentences, however. Such natural sentences land in paradox for the same reasons the more formal $\ulcorner \psi \urcorner$ landed in paradox,

ψ $(\forall i)$ $\ulcorner \psi$ is not true$_i$.$\urcorner$

Namely, they are either not true$_i$ at any level i or true$_n$ at some level n. Assume the former, and then they will be true at level k, where $k > i$ (i.e., they are true 'after all'). Assume the latter, and they should come out false at n. Either way, we have a paradox.

What I've attempted to show, then, is that just as the Strengthened Liar can be uttered in natural language, so too can (some) instances of the Super-Liar. Burge is interested in providing a theory of our ordinary notion of truth. Since I've just shown that there are legitimate Super-Liar candidates in natural language, Burge's theory should be able to apply to them as well.

Unlike many other people working on truth theory, Burge specifically is interested in giving a theory of our ordinary notion of truth. He wants to account for liar sentences as they occur in natural language. Thus, he is left with somewhat of a dilemma. He can either give a comprehensive theory of our ordinary notion of truth or not. If he does, then he needs to be able to give an account of natural Super-Liars like the ones I mentioned above. If he doesn't he still needs to be able to give an account of more formal Super-Liars (like $\ulcorner \psi \urcorner$ above). Either way, the problem of the Super-Liar seems to remain for Burge's contextual theory of truth.

References

Beall, J. (2007). Truth and paradox: A philosophical sketch. In D. Jacquette (Ed.), *The handbook of the philosophy of science: Philosophy of logic* (Vol. 10, pp. 325–410). Elsevier B.V.

Burge, T. (1979). Semantical paradox. *The Journal of Philosophy*, *76*(4), 169–198.

Kripke, S. (1975). Outline of a theory of truth. *The Journal of Philosophy*, *72*(19), 690–716.

Tarski, A. (1933). The concept of truth in formalized languages. In *Logic, semantics, metamathematics* (pp. 152–278). Oxford University Press.

Matt Leonard
Department of Philosophy
University of California, Davis
e-mail: mjleonard@ucdavis.edu
URL: http://philosophy.ucdavis.edu/people/mjleonar

Transparent Intensional Logic
A Challenge

Pavel Materna*

Abstract

The paper presents a brief survey of history and philosophical background of Transparent Intensional Logic (TIL) together with a characteristic of its structure. Some other systems that could be compared as well as some particular results are mentioned and a note on possible future development is added.

Keywords: transparent intensional logic, Tichý, hierarchy of types, construction, procedural isomorphism, anticontextualism, trivialization

1 History

1.1 1968–9

In the year 1968 Pavel Tichý (32), a member of the department of logic (Charles University), published in *Filosofický časopis* (Philosophical Journal) an article "Smysl a procedura" (reprinted as "Sense and Procedure" in Tichý, 2004, pp. 77–92), whose basic importance could not be appreciated then (it did not contain any Marxist-Leninist idea so that no great enthusiasm could have been expected in Czechoslovakia, and no interest in reading Czech articles could have been presupposed abroad). The paper contained two basically important ideas:

A. *The (Fregean) sense of a sentence is a(n abstract) procedure.*

---

*This contribution has been supported by Grant Agency of Czech Republic, projects No P401/10/0792 and P401/10/1279.

> [d]espite the fact that the notion of effective procedure has been used in logic only for the study of the syntax of languages, it is easy to see that, taken in an abstract way, the relation between sentences and procedures is of semantic nature; for sentences are used to record the results of performing particular procedures. (Tichý, 2004, p. 81)

In this article Tichý suggested the way of defining basic semantic categories (truth, synonymy, analyticity) in terms of sense, which in turn is defined as a Turing procedure. This idea has been thoroughly worked up so that in *Studia Logica* next year appeared his paper "Intensions in Terms of Turing Machines" (Tichý, 1969, 2004, pp. 93–110).

> **B.** *Sense cannot be defined "by means of the notion of synonymy or analytical identity of expressions" but conversely synonymy and analyticity are definable in terms of the notion of sense.*

It is erroneously supposed that "the relation of synonymy or analytical identity is definable without the notion of sense." (Tichý, 2004, p. 81).

Interestingly enough, reading the first half of **B**, every follower of Quine's philosophy would approvingly nod his/her head. Actually, the famous arguments from *Two dogmas of empiricism* support the thesis that it is impossible to use notions of synonymy and analyticity as means of defining sense (meaning). The second half of **B** offers however a solution of this problem, a solution, which Quine could not have accepted because *meaning* was an "obscure" notion for him from the very beginning.

Tichý shows that as soon as sense (meaning) is defined the other semantic categories are defined in terms of the definition of sense. In this connection Tichý appreciates the approach of classical logic to defining semantic categories, obviously alluding to Aristotle's theory of definition:

> The sense of a term (in classical terminology rather the "content of concept") is understood as a collection or a family of features, i.e. properties, which is something that does not logically depend on any semantic notion, in particular not on the notion of truth. Just the opposite, the

notion of truth and analytical truth logically depends on the notion of sense:... (Tichý, 2004, p. 81)

Later below it will be clear that the moments characterized by **A** and **B** can be conceived of as a germ of TIL. At least two points are important:

> *Sense (meaning) must be defined independently of (other) semantic categories, which in turn become definable.*

The calling for a "structured meanings" (cf. e.g. Cresswell in 1975, 1985) is satisfied here, first, by *procedural character of sense*, including the Turing-machine realization, second by the manner the classical theory is appreciated.

1.2 1969–1988

The years preceding Tichý's monograph *The Foundations of Frege's Logic* (1988) can be characterized as follows:

a) Development of 1st order TIL, definition of *constructions*,

b) transition from atemporal TIL to a thoroughly elaborated theory of tenses and aspects,

c) essential contributions to many actual problems of logical semantics.

Ad a):

The key notion of TIL, i.e. *construction*, is defined in this period. Constructions are *used* in the particular analyses but they are not *mentioned*: they are not yet objects *sui generis* that can be talked about. This is why TIL is here 1st order only. Type-theoretically, it is based on *simple hierarchy of types*.

Ad b):

In 1980 two important articles appeared: in *Linguistics and Philosophy* ("The logic of temporal discourse") and *Theoretical Linguistics* ("The semantics of episodic verbs"), the former analyzing *tenses* in English, the latter elaborating analysis of *aspects* (in English), i.e. of the cases where events and episodes play essential role. Since then the type of intensions is not only $(\alpha\omega)$: mostly it is $((\alpha\tau)\omega)$, abbreviated as $\alpha_{\tau\omega}$, i.e. intensions are mostly functions from worlds to *chronologies* of some type.

Ad c):

In (Tichý, 2004) practically all Tichý's articles have been scanned and re-edited[1]. Therefrom we can see the scope of topics dealt with by Tichý. The articles VII through XXXVII capture the period that we have called 1st order TIL. We can state that even within this 1st order period Tichý's contributions to solving semantic and philosophical problems are original and at least surprising for many contemporary semanticists. (See, for example, the brilliant criticism of then prevailing (and, among others, from Frege inherited) contextualism in the article "De dicto and de re".)

1.3   1988–

In 1988 Tichý's monograph *The Foundations of Frege's Logic* appeared at de Gruyter. This event can be considered as a milestone in the history of TIL: here a *ramified hierarchy of types* is defined, and since then the higher-order types begin play their role. This point means that the 1st order types fail as soon constructions are *mentioned* rather than *used*. From this moment we can speak about 2nd order or perhaps higher-order TIL. The therewith connected intuition is basically clear and will be dealt with in ch. III.

In this last (final) period Tichý presented (among others) some principles that determine his attempt to realize a great project of "meaning-driven grammar" (see papers XLIII and XLIV in Tichý, 2004)[2] and published two excellent contributions, one from a Wittgensteinian Paris conference 1992 ("The *Tractatus* in the Light of Intensional Logic")[3], the second one (appeared *post mortem* in 1995) from a Vienna conference *Foundational Debate: Complexity and Constructivity in Mathematics and Physics*, where a precise analysis of the re-

---

[1] Those ones which have been originally published in Czech were translated to English.

[2] Unfortunately, Tichý died 1994 and could not finish this project.

[3] The English translation from French: see (Tichý, 2004, pp. 789–800).

lation between functions and constructions is defined ("Constructions as the Subject Matter of Mathematics").

During the second an third period a small group of sympathizers and even followers of TIL has come into being, mostly in Czechoslovakia. Many articles in Czech and Slovak[4] and some books in the same languages have been published, later some articles in English[5] and two books in English (*Concepts and Objects* 1998 in *Acta Philosophica Fennica 63*, *Conceptual Systems* 2004, Logos Verlag, both by Pavel Materna) appeared. A systematic exposition of the contemporary stage of TIL appeared in 2010 (Duží, Jespersen, Materna: *Procedural Semantics for Hyperintensional Logic*).

2 Philosophy

In (Duží et al., 2010) we say that TIL "is an unabashedly Platonist semantics". To understand exactly what is meant thereby we have to read in (Tichý, 1988, p. vii) what *Platonism* (and *Realism*) means to him:

> [p]latonism, the view that over and above material objects, there are also functions, concepts, truth-values, and thoughts.
>
> ... realism, the idea that thoughts are independent of their expression in any language and that each of them is true or false in its own right.

The fact that TIL is "unabashedly" Platonist and Realist in the above sense makes it automatically a heretical system w.r.t. the prevailing more or less nominalistic mainstream, which exploits the neopositivist *horror metaphysicae* and tries to argue that any attempts to do semantics as referring to abstract extra-linguistic entities can be accused of the forbidden pernicious metaphysics. Thus the logical semantics of natural language becomes a kind of empirical (even behaviorist) pragmatics—see Quine's criticism of Carnap—and loses its logical, *a priori* character. In contrast, for TIL the semantic categories are

---

[4]In particular M. Duží, P. Materna, J. Raclavský, P. Kuchyňka, P. Kolář, J. Štěpán, (Slovaks:) P. Cmorej, F. Gahér, M. Zouhar.

[5]M. Duží, P. Materna, B. Jespersen

not definable unless *a priori* relations between expressions and extra-linguistic objects are detected and defined as denoting (the *denotation*) and expressing (the *sense/meaning*).[6] (Because of this view some philosophers call TIL a "Neofregean theory".)

This basically Realist approach influences further points. First, for TIL semantics is logically prior to syntax, at least in the following sense:

> [l]ogic, just like arithmetic and geometry, treats of a specific range of extra-linguistic entities given prior to any axiomatization...

Again, this is incompatible with the dominating syntacticism, for which it is axiomatic formulation what determines semantics.

For TIL, logic studies

> [l]ogical objects (individuals, truth-values, possible worlds, propositions, classes, properties, relations, and the like) and ... ways such objects can be constructed from other such objects. (Tichý, 2004, p. 295)

This is a key quotation. It is not meant as a definition of logic (it would be a circular definition, of course) but as a characteristics that does justice to the philosophical core underlying the approach to logic. Tichý continues to show *via* some simple examples what kind of activities a logician is interested in. A logician is interested in the way a *proposition* that Bill walks can be the result of combining Bill, the *individual*, with the *property* of walkerhood. Similarly, logic will study the way the property of walkerhood combines with other objects to get, e.g., the proposition that everything walks. The resulting *constructions* can detect some interesting relations between objects constructed, for example it can be discovered that the latter proposition *implies* the former proposition. Besides, constructions can be associated with expressions as their 'analyses'—clearly, by 'analyses' Tichý means just *meanings*. (Here: the former construction is the meaning of the sentence "Bill walks", the latter of the sentence "Everything walks".

---

[6]Tichý does not use Frege's terminology (*express—ausdrücken, denote—bezeichnen*). In (Duží et al., 2010) we return to it, though.

Transparent Intensional Logic: A Challenge

This conception of logic is not commonly accepted. The most recent great logicians who would at least partially consent are Frege and Gödel, both proving that realism is not incompatible with exactness.

In (Duží et al., 2010) we write:

> To get your head around TIL, don't think in terms of language-meets-language; think in terms of language-meets-reality. (Duží et al., 2010, p. 56)

What is 'reality' in the case of logic? Considering the above quotation from (Tichý, 2004) and the respective examples we can say that the extra-linguistic reality studied by logicians consists of ways in which particular objects are combined to get other objects. These ways are just *constructions*.

Before we suggest the way constructions are *defined* in TIL (which is a rather technical problem) let us formulate some pre-theoretical intuitions connected with this kind of entity.

First of all: yes, they are *objective, extra-linguistic* entities. This has to be emphasized especially if intuitionists or constructivists come and say, well, we use the notion of constructions as well. Right, but first, intuitionist constructions are specified rather as 'proof-objects'— at least in the case of Martin-Löf—while constructions in TIL get much broader definition, and second, the mentalistic conception of constructions (Brouwer's tradition) is incompatible with the objectual conception defended by TIL. (See, however, Ch. 4.)

(True, but mentalism can be essentially weakened in some intuitionist works. As a good example may serve Fletcher's (1998), where we can read following formulations:

> [A] construction ... is defined recursively as either an atom or $C(x_1, \ldots, x_k)$, where C is a combination rule and x_1, ..., x_k are constructions satisfying the conditions for applying C. ... A *mathematical* construction is an *abstract* recursive structure .... (Fletcher, 1998, p. 51)

This is well compatible with TIL, where constructions are *abstract procedures*.

Besides, Ranta in his (Ranta, 1994) offers a non-mentalistic interpretation of Martin-Löf's Constructive theory of types.)

Further: As soon as abstract procedures (e.g. defined as TIL constructions) become *meanings* of expressions the problem of undefinability of meaning in terms of synonymy and analyticity gets its solution without Quinean famous resignation ("obscure entities") or behaviorist pragmatization of this notion. On the other hand, synonymity and analyticity are definable in terms of meaning. Expressions are synonymous iff they share meaning, and analytically true iff their meaning constructs **T** or the proposition **TRUE** returning **T** in any possible world-time.

Meanings, i.e. constructions, are thus definable independently of the notions of synonymy and analyticity. In TIL, there are strong reasons in favor of the conviction that constructions (and so meanings) cannot be satisfactorily defined (explicated) unless the 1st order (set-theoretical) paradigm is abandoned and higher-order types introduced.

Since constructions are considered to be objective abstract procedures a seemingly controversial or at least strange consequence follows:

Meanings are independent of language. They are not created by language, they are *used* by language. Linguistic convention associates expressions of the given language with meanings. The respective grammar determines the rules of combining simple meanings so that new meanings arise. For example knowing already what the word "black" means we need a rule that would enable us to state that a certain object XY (also known to us) is black. The grammatical rules of predication make it possible to express the fact in English by naming XY and using copula: "XY is black". But undoubtedly neither this fact (of XY being black) nor the possible alternative fact given by the negation thereof is created by language so that the meanings of the respective expressions cannot be freely excogitated entities, they have to somehow correspond to the way in which objective properties and relations are given to us through a given language.[7]

A most important feature of TIL, one which distinguishes TIL from most other theories of meaning, is given just by identifying meanings with constructions: TIL is an *anti-contextualistic*[8] system. *Any expression (of a natural language) has the same meaning independently of the context in which it occurs.* The classical Fregean contextu-

---

[7] We cannot convince each semantic subjectivist, relativist etc. Here we simply refer to some basic ideas underlying TIL.

[8] "transparent" means just this feature.

alism, which has been justified by the famous "reference shift", has been replaced by the "supposition shift". For example, the meaning of the expression "the President of Czech Republic" is the same in both contexts: a) "The President of Czech Republic is a sportsman" b) "XY thinks that the President of Czech Republic is a Slovak", although the former sentence—unlike the latter—does not possess any truth-value as soon as there is no President of Czech Republic. The point is that the meaning (construction!) of "the President of Czech Republic" is used in the *de re* supposition in the case of the sentence a) and in the *de dicto* supposition in the other case. The meaning (construction!) itself remains the same.[9]

A remark concerning philosophical foundations of the theory of constructions: Constructions are abstract procedures. Can we therefore say that constructions are *algorithms*?

The negative answer is justified in Tichý's (1986, 2004, p. 613). Two points distinguish constructions from algorithm:

a) Algorithms solve, in general, mass problems, i.e. they apply to any input from a definite class so that the particular steps differ depending on the particular input. By contrast, constructions always represent a fixed sequence of steps, so they are correlative

> [n]ot with the notion of algorithm itself but with what is known as a particular algorithmic *computation*, the sequence of steps prescribed by the algorithm when it is applied to a particular input.[10]

b) Yet we even cannot claim in general that (all) constructions are algorithmic computations because the latter must contain just *effective* steps. This does not hold for all constructions, which may contain steps that are not effective:

> As distinct from an algorithmic computation, construction is an ideal procedure, not necessarily a mechanical routine for a clerk or a computing machine. (ibidem)

---

[9]Besides—against Frege—"the President of Czech Republic" does *not* denote Václav Klaus in a) and in b) either.

[10]We have to admit, however, that in some cases the construction behaves as an algorithm (with the proviso sub b)), e.g. when the former is an open construction, i.e. when it contains free variables.

(In 1968, 1969, when Tichý showed that meaning can be explicated in terms of Turing machines, he already took into account this distinction: explicating meaning of an empirical expression he used the *oracle* box, so *O*-machines, for which Church-Turing Thesis does not hold.)

A question may arise: Is TIL a *classical* logic?

There are many attempts at distinguishing classical logic and non-classical logics. One criterion is given by the *principle of bivalence*, to which the classical logic does and (some) non-classical logics do not adhere. There are however two options how to define the principle of bivalence:

a) *Every (declarative) sentence is either true or false.*

b) *There are just two truth-values.*

For TIL the first formulation is not acceptable: since functions are treated as *partial functions* there are sentences that lack any truth-value (observe, e.g., "The greatest prime is odd"). Thus we could be tempted to say that TIL is a many-valued logic. Yet TIL adheres to the principle of bivalence in the second sense: There are just two truth-values, say, **T** and **F**. Thus some sentences simply lack any truth-value. The absence of a truth-value is not some 'third' truth-value. So the matrices of truth-values differ from Łukasiewicz-like matrices. (They could be compared rather with Boczvar's matrices, where—similarly as in TIL—the occurrence of one 'third value' induces this 'third value' of the whole construction. Thus the sentence "The greatest prime is odd or the smallest prime is even" gets the value **T** for Łukasiewicz and the 'third value' for Boczvar as well as no value for TIL).

Further, TIL offers a unitary *approach* to solving logical problems. For example, the semantics of interrogative sentences can be analyzed without building up separate 'erotetic logics', TIL can treat modalities so that particular modal logics get particular justifications (and they themselves solve purely formal problems concerning, e.g., properties of particular formalizations), epistemic and doxastic logics get or can get deeper semantic foundations etc. Problems that are dealt with in fuzzy logics can be solved using TIL, a deeper discussion with intuitionists can be realized etc.

Not everything has been accomplished in this respect (and it will never be, of course), but I am talking about *ambitions* of TIL, and these ambitions make of TIL an enemy of plurality of logics. *There are various ways how to develop logic, and the particular way being realized by TIL may be just one of them*[11], *but this viewpoint does not justify the view that there can be various logics.*

From this viewpoint I would rather admit that TIL is one of the ways how to develop the *classical logic*. There are some points where we can agree with Bocheński's conception of logic (see Banks, 1950), for example with his emphasis on *rules* and his statement that no particular formal system of a non-classical logic has got a non-classical system of metalogical rules.

3 Architecture: 3 levels

Hyperintensionality is sometimes defined or at least characterized negatively, i.e. as a property of those texts which cannot be analyzed (not only in terms of extensionalist but even) in terms of intensionalist principles. Thus Cresswell says in (Cresswell, 1975, p. 25) "Hyperintensional contexts are simply contexts which do not respect logical equivalence.", or Williamson

> An operator O is hyperintensional iff it is not non-hyperintensional, and O is non-hyperintensional iff the following condition is satisfied:
>
> > If p is strictly equivalent to q, then Op is strictly equivalent to Oq.
>
> (Quotation from a blog devoted to issues in language, epistemology, metaphysics, and mind by Lemmings)

(Cf. also, e.g., Bealer in 1982, where "qualities", "connections" and "conditions" are intensions in the sense of "the first conception" while "concepts" and "thoughts" are "the other intensional entities". For the former it holds that they are "identical if and only if they are necessarily equivalent", while each definable intensional entity of the latter kind "is such that, when it is defined completely, it has a unique, non-circular definition". (Bealer, 1982, p. 2)

Elsewhere Bealer (i.e., in 1989, p. 10) says:

---

[11] For example, we can imagine a similar system based on Curry's combinators instead on λ-calculus.

> How is one to develop a theory of the *other* type of intension? This job will require some new kind of logical machinery, machinery not used in the original propositional-function approach...

Bealer is obviously not content with a negative determination of hyperintensions. His solution consists in an algebraic approach, which means that (at least in Bealer's realization) the problem of predication of empirical attributes is not satisfactorily solved while the direct approach in TIL, which uses *variables for possible worlds and time moments* (this is called *explicit intensionalization*) is able to explicitly distinguish between empirical and non-empirical expressions, which is highly relevant from the logical viewpoint.

Bealer's algebraic approach is characteristic of the way most logicians choose to analyze natural language. This way (classically exploited by Montague) is described and confronted with TIL in (Duží et al., 2010):

> The contemporary mainstream method of logically analyzing expressions of a natural language consists in building up an artificial language and defining some rules of translation that make it possible to find for every expression of the given language its translated counterpart in the artificial language. The latter is unambiguous (unlike the former) and is interpreted in a model in the usual way. Tichý calls this method *formalization*. Formalization itself, if thought of as a means to make ideas precise, is indispensable. The method deployed by TIL to make ideas precise is [however] a method of *direct* analysis. The notion of construction enables us to justify this direct transition from expressions to their meanings. (Duží et al., 2010, p. 95)

The transition to hyperintensionality is therefore realized *via* defining types and constructions of higher order.

The resulting hierarchy of types consists of **A.** *types of order 1* and **B.** *higher-order types*. In brief:

A. The *basic* types[12], here:

ι ...... Universe, individuals   compare Montague's e

o ...... $\{\mathbf{T}, \mathbf{F}\}$, truth-values   compare Montague's t

τ ...... real numbers / time moments

ω ...... logical space, possible worlds

The functional types: $(\alpha\beta_1 \ldots \beta_m)$, sets of partial functions with α the type of the values, β_i, types of arguments.

(So the types of *intensions* are (for α any type) $((\alpha\tau)\omega)$, $\alpha_{\tau\omega}$ for short.)

Before defining the higher-order types we have to define *constructions*. As for exact definitions and more details see (Tichý, 1988) and (Duží et al., 2010).

Variables. For any type countably infinitely many variables are at our disposal. They are a kind of (extra-linguistic) constructions (the respective letters are just names of variables), which construct objects of the given type dependently on a total function *valuation*, so they v-construct objects, where v is the parameter of valuations.

Trivialization. If X is an object, then ^0X is a construction called *trivialization*. It *mentions* X and lets it be without any change. This construction is highly important—it makes it possible to 'jump' to the higher-order types and thus to hyperintensionality.

Execution and **Double execution:** we will not need to use them in the present general exposition.

The last two constructions are objectual counterparts of λ-terms. It is not by chance that the respective λ-terms are used by Montague as well: Montague grasped the ingenious Church's idea of essential reducibility of (most? All?) operations to *applying functions to arguments* and *creating functions by abstraction*.

---

[12]The choice of *base* is dependent on the language that has to be analyzed. The present choice proved its expressive power in application to a natural language.

Composition. Where $X, X_1, \ldots, X_m$ are constructions and X (v-)constructs a *function* F (thus of type $(\alpha\beta_1 \ldots \beta_m)$), and $X_1 \ldots X_m$ (v-)construct objects of the types $\beta_1, \ldots, \beta_m$, the construction $[XX_1 \ldots X_m]$, called *Composition*, which corresponds to the λ-term *Application*, (v-)constructs the value of F (if any) at the arguments (v-)constructed by X_1, $\ldots, X_m$.[13] Partiality of functions causes that Composition can be (v-)*improper*, i.e. it can construct nothing (like in the case of dividing by zero). Observe that objects that are talked about are never present in Composition (neither in any construction with the exception of Trivialization): they are always represented by a construction. Thus the Composition that constructs the result of dividing 5 by 3 is not $[: 5\ 3]$ but $[^0: {}^05\ {}^03]$.

Closure. Where $x_1, \ldots, x_m$ are pairwise distinct variables v-constructing objects of types (not necessarily distinct) β_1, $\ldots, \beta_m$ and X is a construction v-constructing objects of type α, the construction $[\lambda x_1 \ldots x_m X]$[14], called *Closure*, which corresponds to the λ-term *Abstraction*, v-constructs a function F of type $(\alpha\beta_1 \ldots \beta_m)$. The value of F (if any) at the arguments $b_1, \ldots, b_m$ equals the object (if any) v'-constructed by X, where v' associates $x_1, \ldots, x_m$ with b_1, $\ldots, b_m$, respectively.

Observe that Closure is never (v-)improper: it always constructs a function (maybe a degenerated one, i.e., undefined at each argument, like $\boldsymbol{\lambda x_1}[^0: \boldsymbol{x_1}\ {}^0\boldsymbol{0}]$.

B. Higher-order types are generated by a *ramified hierarchy*.

The idea can be suggested as follows:

i) *Types of order 1* have been defined already (see here **A.**).

ii) *Constructions of order n* are defined: the idea consists in claiming that if a construction C constructs an object of a type of order n then C is *a construction of order n*.

---

[13] Now we will write "construct" etc., meaning that v-construct can be admitted.
[14] The outmost brackets can be omitted.

iii) *Higher-order types* are defined as follows:
Let $*_n$ be the collection of constructions of order n. (See ii).) Then $*_n$ and every type of order n is a type of order $n + 1$. (And if the types $\alpha, \beta_1, \ldots, \beta_m$ are types of order $n + 1$ then the functional type $(\alpha\beta_1 \ldots \beta_m)$ is a type of order $n + 1$.)

Thus the transition to hyperintensionality has been accomplished.

Observe that this transition is based on the distinction between *using* and *mentioning* constructions. In $[^0{=}[^0{+}\,{^03}\,{^05}][^0{:}\,{^056}\,{^07}]]$ the two subconstructions whose equivalence is claimed are *used*: the identity concerns the objects constructed by them. On the other hand, this identity does not justify the claim that if XY calculates $3 + 5$ then XY calculates $56 : 7$. The respective constructions, i.e.

$$\lambda w \lambda t [^0\text{Calculate}_{wt}\ {^0}\text{XY}\ {^0}[^0{+}\ {^03}\ {^05}]],$$
$$\lambda w \lambda t [^0\text{Calculate}_{wt}\ {^0}\text{XY}\ {^0}[^0{:}\ {^0}56\ {^0}7]]$$

concern the respective subconstructions themselves rather than what they construct: they are *mentioned*, trivialized. (The type of Calculate is not $(o\iota\tau)_{\tau\omega}$, it is $(o\iota*_1)_{\tau\omega}$.) Leibniz's rule does hold but is not applicable.

As for *using* vs. *mentioning* we can distinguish three levels, see Figure 1.

This "three-levels schema" makes it possible to make more semantic distinctions than can be made in other theories, which are satisfied with at most functional level (Kripke, Montague, ...). An important point should be emphasized: within this schema compositionality is fully observed, and the cases where semantic distinction is stated between logically equivalent expressions are explained without flouting any principle of extensionality. So we can state that TIL is an extensional (albeit hyperintensional) logic.

To exemplify our claim (possibility of making more semantic distinctions...) let us analyze the following argument:

$$3 + 5 = 56 : 7$$
$$\frac{\text{Charles calculates } 3 + 5}{\text{Charles calculates } 56 : 7}$$

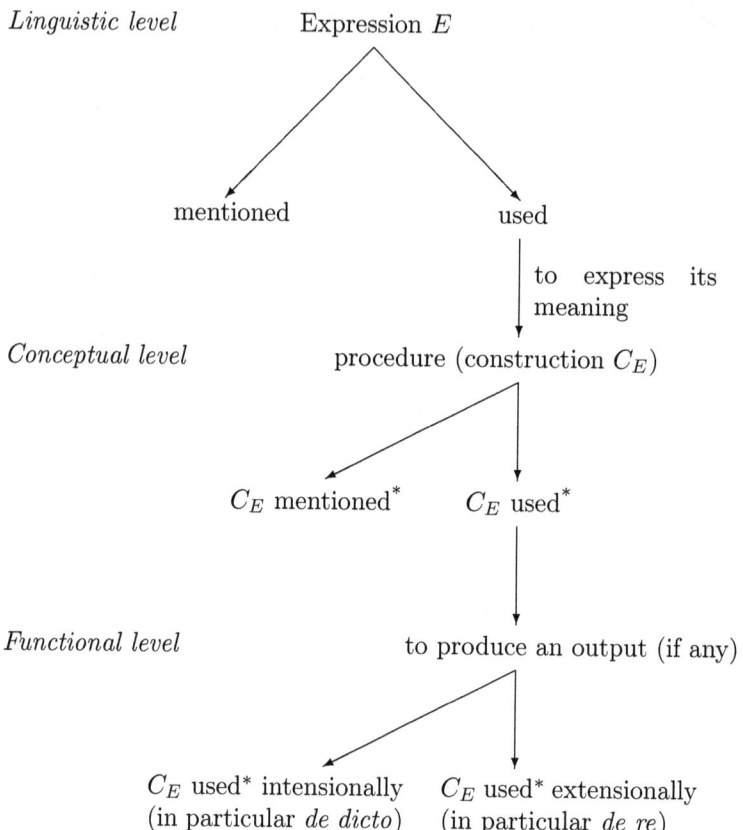

Figure 1: *Using* vs. *mentioning*.

It seems as if the conclusion were justified by Leibniz' Substitution rule but as soon as premises are true the conclusion is evidently false. Does it mean that Leibniz' rule should be revised?

Surely not. Let us analyze the premises.

Types: $+/(\tau\tau\tau), :/(\tau\tau\tau), =/(o\tau\tau), 3/\tau, 5/\tau, 56/\tau, 7/\tau$, Charles$/\iota$, calculate$/(o\iota *_1)_{\tau\omega}$

$$[^0= [^0+ {}^03 {}^05][^0: {}^056 {}^07]]$$

$$\lambda w \lambda t[^0\text{Calc}_{wt}{}^0\text{Ch } ^0[^0+ {}^03 {}^05]]$$

We can see that Leibniz holds but *cannot be applied*: the identity concerns what the two constructions *construct*, while Charles is related to one of these constructions rather than to its value. (The construction is *mentioned* in the second premise.)

4 Comparable theories

There are logical/semantic theories which are simply incomparable with TIL. It would be not very productive to try to compare TIL, e.g., with Quine or the later Wittgenstein, to confront TIL with inferentialism and the like. The reason is that all such (similar) streams and trends are based on philosophies incompatible with (that of) TIL (see ch. 2 here).

On the other hand, some (realist) theories can be compared with TIL since they do not try to replace semantics by pragmatics. They attempt at more fine-grained solutions of semantic problems than those ones used by current 1st order theories. Besides, among the non-realist theories that are in some respects comparable with TIL intuitionist/constructivist theories have to be mentioned, in particular the Constructivist Type Theory (CTT) by Per Martin-Löf.

4.1 Realists

TIL can serve as a *logical analysis of natural language* (LANL). The most popular version of LANL descends however from *Richard Montague* (1974). Montague's logic shares at least two features with TIL: it uses a *functional language* and its universe is *type-theoretically* classified. There are however some important features of Montague's approach that can be criticized, which Tichý did in (Tichý, 1988, 2004, in particular pp. 831–835).

In the paragraph 2.4.3. of (Duží et al., 2010) there are adduced and criticized some features of Montague's logic IL and TY2 which are distinct from TIL, in particular absence of Church-Rosser 'diamond property' (Montague, like most logicians, does not use explicit intensionalization—see above), confinement to total unary functions (Schönfinkel's reduction to unary functions holds only for total functions), and, what is probably most important, the fact that Montague cannot analyze attitudes that concern non-empirical concepts: such attitudes relate individuals with (structured) complexes (but not with linguistic expressions). In TIL this means that such attitudes relate individuals with constructions. (Remember that, e.g., believing that $2+2=4$ differs from believing that arithmetic of naturals is incomplete.) Montague does not realize the 'jump' to hyperintensionality. Therefore, some skepticism concerning the ability of semantics of natural language arose together with the opinion that some problems have to be solved by pragmatics. See (Gamut, 1991, pp. 73–74):

> The relations between language and language users can to a large extent be abstracted away in semantics, but not entirely, and the analysis of belief contexts is thought to be one area in which the semantic interpretation must take language users into account.

Another important theory is the Neo-Meinongian theory by Edward Zalta (see, e.g., his 1988a, 1988b). Zalta's diagnosis of Montague's failure is exact. He says about Montague's reconstruction of relations and propositions:

> [s]uch a reconstruction doesn't capture one of their most important features, namely, that two relations or propositions may be distinct even though necessarily equivalent. ... [consequence] that the semantic value of the parts of the complex expression cannot be recovered from the semantic value of the entire expression. (Zalta, 1988a, pp. 62–63)

Briefly, Montague's system does not capture structured meanings.

It's not clear whether Zalta's therapy works. His main means are: distinguishing between *encoding* and *exemplifying* (which could be compared with *intensional* and *extensional use*[*] of constructions in

TIL)[15] and defining "logical operations" like '$PLUG_i$' (that inserts objects into i-ary relations). The purpose is clear: if expressions denote such 'pluggings' into relations then the denotation is *complex*, containing the plugging operation plus the denotation of the respective relation constant and denotations of the respective individual constants. Some comments are however advisable.

The logical character of 'pluggings' is not clear enough: They are functions (as mappings). On this interpretation the result would not correspond to Zalta's aim because functions are insensitive to structures. True, as defined on *expressions* we get a 'more complex' expression but the object denoted is a set-theoretical, i.e. a simple object.

As for TIL, no such extra operation of plugging is needed. The application of a function to arguments is sufficient, and since the meanings of two logically equivalent but distinct expressions are distinct *constructions* Zalta's purpose is achieved.

Zalta's 'pluggings' remind us of a similar attempt at capturing the logical character of predication in Bealer's PRP theory (1979)—or see (Bealer & Mönnich, 1984)—with his 'pred' function. See (Duží et al., 2010, pp. 196–200).

One point in Zalta's approach makes TIL closer to Montague than to Zalta: The latter says:

> [a] type Theory based on relations and relational application is slightly more general than a type theory based on functions and functional application (since functions are a special case of relations) (Zalta, 1988a, p. 85)

First, Zalta obviously forgot that relations are a special case of functions as well. Second, functions are preferred by Montague and by TIL because they are—unlike relations—'construction-friendly': they can be *applied to* arguments (Composition!), which is one of the most important procedures. Third, working with partial functions we can distinguish the case where the function does not apply to the argument from the case where the function is *undefined* at the argument. This distinction is not detectable if only relations are taken into account.

Independently of TIL, the idea of sense/meaning as a procedure/algorithm has been explicitly formulated and (in the spirit of Montague) elaborated in the works of *Y. N. Moschovakis* (1994, 2006). It was

---

[15] A good comparison of Zalta with Tichý can be found in (Sierszulska, 2006).

(not only but notably) Moschovakis, whose idea inspired *Reinhard Muskens* (e.g. in his 2005) to elaborate on "the Fregean idea that the sense of an expression essentially is a method or algorithm" (Muskens, 2005, p. 473). Muskens shows that Thomason's study (1980) can be reinterpreted in a computational way, and his method is obviously more or less applicable to other semantic systems, where we can distinguish expressions concerning three domains: an *algebraic* domain, say, **A**, that represents logical notions, domain of *propositions*, say, **P**, and the *connecting* domain, say, **C**, which associates **A** and **P**. Hyperintensionality (in the sense "intensions are not sufficient") is reached due to non-identity of **A** and **P**: If **A** were identical with **P** then propositions would be definable within **A** and the sustainable cases where logically equivalent expressions are semantically distinct could not be explained. To hold **A** and **P** distinct we need the domain **C** that establishes the connection between them.[16] Muskens' idea consists in "giving a computational interpretation to the connection **C**" (Muskens, 2005, p. 486). The respective axiomatization is realized by means of a (Prolog-like) logic program. Propositions are interpreted as queries.

> Distinct queries can lead to the same result and identity criteria on queries can be very strict, thus leading to intentional (hyperintensional) semantics. (Muskens, 2005, p. 502)

Muskens' sympathetic attempt to connect the notion of meaning with a kind of program differs from TIL though. The choice of a particular logic program has been made with the view of enabling us to make the results of some queries diverge, while TIL defines a natural level of mentioning procedures/constructions, so that a category of constructions becomes a selfcontained kind of entities. The respective rules lead to hyperintensionality in a natural way (and more generally than in Muskens). It turns out also that the method of explicit intenzionalization (see above) is a very useful method that exploits conveniences connected with manipulating variables and that the way TIL solves the problems with *de re* vs. *de dicto* is a general way based among other things on this method without using the controversial 'functors'

---

[16]Another way of making **A** and **P** distinct has been proposed by Hintikka (1975) and applied by several not only paraconsistent logicians: if possible worlds are not sufficient, add 'impossible worlds'. This step is in principle alien to TIL.

∧ and ∨.[17] After all, neither Montague nor Muskens is able to solve the cases where procedures are mentioned (like in Charles' calculating $3 + 5$ vs. $3 + 5 = 56 : 7$) (see above ch. 3, **B**).

4.2 Intuitionists

Intuitionists share with TIL the idea of *construction*. Sometimes this expression is explicated by intuitionists so that the similarity with TIL is remarkable. So we can read in (Fletcher, 1998):

> If one had to define constructions in general, one would surely say that a type of construction is specified by some *atoms* and some *combination rules* of the form "Given constructions $x_1, \ldots, x_k$ one may form the construction $C(x_1, \ldots, x_k)$, subject to certain conditions on $x_1, \ldots, x_k$"
> (Fletcher, 1998, p. 51)

Fletcher's book is a most interesting exposition of one sort of intuitionism which is not strongly dependent on Brouwer's mentalistic conception of mathematics. Reading it one cannot avoid the idea that there are some interesting points that are shared by TIL and intuitionism. In (Primiero & Jespersen, 2010) the authors adduce following "common features":

- a notion of construction;
- a functional language;
- a typed universe;
- an interpreted syntax.

The authors believe that a "neutral notion of procedural semantics" is definable due to the above shared features and in the text that follows this conjecture they adduce interesting arguments.[18] In my opinion some further research is needed, since only a partial harmony is demonstrated in the article, concerning the problems with analyzing

---

[17]This remark concerns rather the Montagovian school. Muskens at least uses "$\lambda i \ldots$" for i ranging over possible worlds.

[18]Intuitionism is here represented by the approach known from the works by *Per Martin-Löf*. In particular, the system TIL is here compared with is CTT (Constructive Type Theory).

privative modification. Theoretically some problems might appear which would be solvable only by one of the 'competing' theories. Also: The theory of proofs as constructions does not lack some weak points in CTT: *empirical* concepts can hardly be given by "empirical proofs", at least in the recent stage of development of CTT. It seems however that some new research could succeed in this respect (Primiero & Jespersen, 2010). Besides, the possibility of non-mentalistic proofs of non-mathematical propositions within the constructivist frame has been demonstrated by Aarne Ranta (1994, §2.26).

5 Particular results

Now we will mention some more or less general results of logical analyses made in terms of TIL. The following list is not exhaustive and it is not ordered according to some clear criterion as, e.g., generality of the respective problem, size of the solution, originality etc.

1. Rules of hyperintensional partial typed λ-calculus

    Extensionality

    Properness

    Improperness

    Existence

    Substitution:

    > The extensional rule
    > The intensional rule
    > The hyperintensional rule

2. General definition of *de re* vs. *de dicto*

    > De re supposition is defined as a particular case of extensional supposition, de dicto supposition as a particular case of intensional supposition.

3. A fine-grained analysis of tenses and aspects in English

    > Simple past, Present perfect, temporal *de dicto* vs. *de re*, Future tenses

4. A fine-grained hyperintensional theory of propositional and notional attitudes

   Implicit (intensional), explicit (hyperintensional) and inferable knowledge

5. A general theory of identities

   Including the "Hesperus-Phosphorus" problem

6. Theory of requisites, intensional essentialism

   Individual essentialism rejected

7. Theory of property modifications

   Rules of pseudo-detachment

8. Anaphora, donkey sentences

   Pre-processing the anaphoric reference

9. Proof of validity of β-reduction 'by value'

   In general, β-reduction is not an equivalent transformation.

10. Proof of hyperintensional validity of the Compensation Principle

    A generalization of Tichý's proof for 1st order TIL

11. Analytically vs. logically true sentences

    Analytic vs. semantic information

12. Procedural theory of concepts

    Empty concepts, empirical concepts, conceptual systems

13. Analysis of questions

    Semantic core vs. pragmatic moment

14. Modalities. Logical vs. nomological necessity

    Counterfactuals

The notably philosophical character of the following results:

15. Anti-contextualist definition of meaning, preserving compositionality

16. An original explication of possible worlds

17. Classification of properties

18. A thorough argumentation for anti-actualism, connected with criticism of Dummett and Kripke

6 Perspectives

TIL is "an open-ended theory with a cast-iron core". (See Duží et al., 2010, p. vii) The 'cast-iron core' consists of the fundamental principles that define the approach to analyzing expressions and make up (together) the essence of TIL. They might be called '*TIL dogmata*'.

'open-ended' means: a) TIL can be extended, attempting at solutions to further problems, b) we admit that particular errors can be found in the theory: these should be corrigible without changing 'dogmata', c) some new 'dogmata' can be added, i.e. some theorems can turn out to be most important from the viewpoint of the 'old' dogmata.

In what follows we adduce some topics that could be dealt with during some further development of TIL.

- Solution to some actual semantic problems

- Hyperintensional generalization of Tichý's calculus for the 1st order TIL

- Development of the computational version of TIL (TIL-*Script*)

- Problems with vagueness

- Space-time type instead of τ

- LANL based on Curry's combinators

- Paradoxes (Raclavský)

- Norms

Concluding remark

In my opinion plurality of most 'specific' logics can be explained as follows: there is no unified approach to logical analyses (logic develops mostly 'bottom-up'). TIL is one of the options how to proceed in solving particular problems of philosophical logic and LANL. (See our approach to analyzing questions.) It develops 'top-down'. Thus what TIL offers is to formulate problems as follows:

> *Logical character of* norms
>> vagueness
>>
>> tenses
>>
>> imperatives
>>
>> etc.

instead of

> *Build up an axiomatic system of* the logic of norms
>> the logic of vagueness
>>
>> the logic of tenses (temporal logic)
>>
>> the logic of imperatives

TIL is in principle able to proceed in this way because its hyperintensional character is connected with high expressivity that makes it possible to choose the top-down development (see Duží et al., 2010, pp. 35–37).

References

Banks, P. (1950). On the philosophical interpretation of logic: An Aristotelian dialogue. *Dominican Studies*, *3*(2), 139–153.

Bealer, G. (1979). Theories of properties, relations, and propositions. *Journal of Philosophy*, *76*, 634–648.

Bealer, G. (1982). *Quality and concept.* Oxford: Clarendon Press.

Bealer, G. (1989). On the identification of properties and propositional functions. *Linguistics and Philosophy*, *12*, 1–14.

Bealer, G., & Mönnich, U. (1984). Property theories. In D. Gabbay & F. Günthner (Eds.), *Handbook of philosophical logic* (Vol. IV). Dordrecht: D. Seidel.

Cresswell, M. (1975). Hyperintensional logic. *Studia Logica, 34*, 25–38.

Cresswell, M. (1985). *Structured meanings*. Cambridge: MIT Press.

Duží, M., Jespersen, B., & Materna, P. (2010). *Procedura semantics for hyperintensional logic*. Springer.

Fletcher, P. (1998). *Truth, proof and infinity*. Dordrecht: Kluwer.

Gamut, L. T. F. (1991). *Logic, language and meaning* (Vol. II). Chicago, London: The University of Chicago Press.

Hintikka, J. (1975). Impossible possible worlds vindicated. *Journal of Philosophical Logic, 4*, 474–485.

Materna, P. (1998). Concepts and objects. *Acta Philosophica Fennica, 63*.

Materna, P. (2004). *Conceptual systems*. Berlin: Logos.

Montague, R. (1974). *Formal philosophy: Selected papers of R. Montague* . (R. Thomason, Ed.). New Haven: Yale University Press.

Moschovakis, Y. (1994). Sense and denotation as algorithm and value. In J. Väänänen & J. Oikkonen (Eds.), *Lecture notes in logic* (Vol. 2, pp. 210–249). Berlin: Springer.

Moschovakis, Y. (2006). A logical calculus of meaning and synonymy. *Linguistics and Philosophy, 29*, 27–89.

Muskens, R. (2005). Sense and the computation of reference. *Linguistics and Philosophy, 28*, 473–504.

Primiero, G., & Jespersen, B. (2010). Two kinds of procedural semantics for privative modification. In K. Nakakoji, Y. Murakami, & E. McCready (Eds.), *New frontiers in artificial intelligence: JSAI-isAI 2009 workshops* (Vol. 6284, pp. 252–271). Berlin: Springer Verlag.

Ranta, A. (1994). *Type-theoretical grammar*. Oxford: Clarendon Press.

Sierszulska, A. (2006). On Tichý's determiners and Zalta's abstract objects. *Axiomathes, 16*, 486–498.

Thomason, R. (1980). A model theory for propositional attitudes. *Linguistics And Philosophy, 4*, 47–70.

Tichý, P. (1968). Smysl a procedura. *Filosofický časopis, 16*, 222–232. (Translated as 'Sense and procedure' in Tichý, 2004, pp. 77–92)

Tichý, P. (1969). Intensions in terms of Turing machines. *Studia Logica*, *26*, 7–25. (Reprinted in Tichý, 2004, pp. 93–109)

Tichý, P. (1980a). The logic of temporal discourse. *Linguistics and Philosophy*, *3*, 343–369. (Reprinted in Tichý, 2004, pp. 373–369)

Tichý, P. (1980b). The semantics of episodic verbs. *Theoretical Linguistics*, *7*, 263–296. (Reprinted in Tichý, 2004, pp. 411–446)

Tichý, P. (1986). Constructions. *Philosophy of Science*, *53*, 514–534. (Reprinted in Tichý, 2004, pp. 599–621)

Tichý, P. (1988). *The foundations of Frege's logic*. Berlin, New York: De Gruyter.

Tichý, P. (1995). Constructions as the subject-matter of mathematics. In W. DePauli-Schimanovich, E. Köhler, & F. Stadler (Eds.), *The foundational debate: Complexity and constructivity in mathematics and physics* (pp. 175–185). Dordrecht, Boston, London, and Vienna: Kluwer. (Reprinted in Tichý, 2004, pp. 873–885)

Tichý, P. (2004). *Collected papers in logic and philosophy* (V. Svoboda, B. Jespersen, & C. Cheyne, Eds.). Prague: Filosofia, Czech Academy of Sciences; Dunedin: University of Otago.

Zalta, E. (1988a). A comparison of two intensional logics. *Linguistics and Philosophy*, *11*, 59–89.

Zalta, E. (1988b). *Intensional logic and the metaphysics of intentionality*. Cambridge, London: MIT Press.

Pavel Materna
Institute of Philosophy, Academy of Sciences of the Czech Republic
Jilská 1, 110 00 Prague 1, Czech Republic
e-mail: `maternapavel@seznam.cz`

Sceptical and Credulous Approach to Deductive Argumentation

Svatopluk Nevrkla[*]

Abstract

I present a deductive argumentation framework of Besnard and Hunter for dealing with classically inconsistent sets. Besnard and Hunter define the notions of an argument and of a binary relation of defeating among arguments. Then, they proceed to define canonical defeaters to a certain argument, which are supposed to be the representatives of all possible defeaters to this argument, that we need to take in account. Further, they proceed to define an argument tree as such a tree, whose nodes are arguments and in which children of each node are exactly all canonical defeaters of the parent, which introduce new premises. They use argument trees to decide which arguments are warranted with respect to some set of beliefs. Finally, they suggest several categoriser and accumulator functions to evaluate members of this belief set on basis of argument trees for arguments supporting or disproving these beliefs. I demonstrate on several examples, that the decision not to include other but canonical defeaters into argument trees affects the resulting values of accumulator functions. I conclude that canonical arguments cannot be treated as sufficient representatives of all arguments.

Keywords: deductive argumentation, canonical counterarguments, argumentation trees, argument evaluation

---

[*]Production of this paper was supported by project GAČR 401/09/H007 "Logical foundations of semantics".

1 Introduction

In past two decades, formal approaches to the study of argumentation have been flourishing significantly in artificial intelligence.

This is mainly due to the landmark work of P.M. Dung on the abstract argumentation (Bondarenko, Dung, Kowalski, & Toni, 1997), but different approaches have developed since Dung's paper, such as assumption-based argumentation (Dung, 1995) or deductive argumentation (Besnard & Hunter, 2001), I examine in this paper.

An exhausting historical overview of formal approaches to argumentation can be found in (Bench-Capon & Dunne, 2007).

Formal argumentation is being applied to such areas of artificial intelligence such as decision making or dealing with inconsistent information in multi-agent environments while also arousing theoretical interest of logicians, mathematicians and computer scientists.

These new achievements in formalization of argumentation also ignite the interest of researchers in disciplines such as informal logic, offering them new perspectives and approaches to the topics of an argumentation and a dialogue in a natural language.

The theory of natural argumentation, as it seems, may benefit from argumentation formalisms in a similar manner epistemology benefits from modal logics or the theory of communication benefits from dynamic logics and the theories of belief change.

Purpose of this paper is to present formal approach to deductive argumentation, first introduced by Besnard and Hunter (2001), which rethinks the role of classical propositional logic for argumentation.

Despite introducing most fundamental concepts like: 'deductive argument', 'counterargument', 'canonical argument', 'argument tree' and 'warranted argument', the authors did not directly address some of the purely formal implications of theirs definitions, but rather motivated them by giving examples of intended application of their formalisms to the natural language argumentation.

Besnard and Hunter introduce the notion of canonical undercut and claim: 'Canonical undercuts are particularly important proposal for ensuring that all the relevant undercuts for an argument are presented, thereby ensuring that a constellation of arguments and counterarguments is exhaustive, and yet ensuring that redundant arguments are avoided from this presentation.'

It is not clear what notions of exsaustiveness and redundancy do authors have in mind, because there is no formal proof of such claim. I will demonstrate, that under certain intuitive interpretation of these concepts, this claim is incorrect due to the restriction on repetition of arguments in argument trees.

2 Deductive argumentation framework

Let's delve directly into definitions of (Besnard & Hunter, 2001).

Definition 1 A (deductive) argument of classical propositional logic (CPL) over a set Δ of propositional formulas (PFs) is an ordered pair $\mathbf{a} = \langle \Phi, \phi \rangle$, where $\Phi \subseteq \Delta$ is a set of PFs and ϕ a single PF, such that:

(1) $\Phi \vdash \phi$.

(2) Φ is consistent.

(3) Φ is a minimal set, such that (1) and (2) hold.

Whenever $\mathbf{a} = \langle \Phi, \phi \rangle$ is an argument, Φ is called the support of $\mathbf{a}$ (Pre($\mathbf{a}$)), while ϕ is called the claim (or conclusion) of $\mathbf{a}$ (Con($\mathbf{a}$)).[1]

Whenever Δ is a set of PFs, the set of all arguments of CPL over Δ will be denoted $[\Delta]$.

While the arguments of other than classical logic could be considered and necessity of conditions (2) and (3) examined, I shall strictly keep this setup.

My 'philosophy' is, that while choosing the details of elementary definitions, we should be guided not only by our intuitive understanding of the field of human knowledge we wish to formalize, or practical utility of our theory, as Besnard and Hunter do, but should also be evaluated from the perspective of nontriviality and mathematical intricacy of the theory that arises.

I will hopefully demonstrate, that the theory developed by Besnard and Hunter is not only interesting as a prospective tool to be used in the computer science, but is also a fruitful source of interesting mathematicial problems. Let us continue further.

---

[1] It follows from requirement (2) and compactness of CPL, that for all arguments the set of premises is finite.

Definition 2 An argument $\mathsf{a} = \langle \Psi, \psi \rangle$ is said to be a defeater of an argument $\mathsf{b} = \langle \Phi, \phi \rangle$ iff $\psi = \neg \bigwedge \Theta$, where $\Theta \subseteq \Psi$.

Besnard and Hunter further introduce conservativity ordering among arguments.

Definition 3 An argument $\mathsf{a} = \langle \Psi, \psi \rangle$ is more conservative than the argument $\mathsf{b} = \langle \Phi, \phi \rangle$ ($\mathsf{a} \triangleleft \mathsf{b}$) iff $\Psi \subseteq \Phi$ and $\psi \vdash \phi$. Arguments a and b are equivalent iff $\psi \equiv \phi$ and $\Psi \equiv \Phi$ (Meaning that the set of valuations satisfying Ψ is exactly the same set of valuations satisfying Φ).

The motivation for introducing this ordering of arguments is expressed by following remark, although not stated explicitly in (Besnard & Hunter, 2001).

Remark In CPL it holds that for each pair of arguments a and b, such that ($\mathsf{a} \triangleleft \mathsf{b}$) and for each argument c it holds: If c is a defeater of a, then c is also a defeater of b. If c is defeated by b, then c is also defeated by a.

Besnard and Hunter prove, that the most conservative defeaters of an argument $\langle \Phi, \phi \rangle$ are equivalent to the arguments of a form $\langle \Psi, \neg \bigwedge \Phi \rangle$. Given a certain canonical enumeration of sentences of our language, we may fix such arguments of this form, that the conjuncts in $\bigwedge \Phi$ are sorted by this enumeration. Such arguments are called canonical defeaters. They further prove that for each argument a and it's defeater b there is a canonical defeater of a, an argument c, which is more conservative than b.

This fact, combined with the previous remark might suggest, that the canonical defeaters truly are the only defeaters we might want to consider, but there is a catch. The applications of the previous concepts in another definitions of (Besnard & Hunter, 2001) will reveal a certain important feature of noncanonical arguments, which canonical arguments do not posses. Let's continue with our exposition of deductive argumentation by presenting these definitions.

Definition 4 An argument tree for $\mathsf{a} \in [\Delta]$ is a tree T consisting of arguments, such that:

(1) The root is an argument a.

(2) For no node $\langle \Psi, \psi \rangle$ with ancestor nodes $\langle \Psi_1, \psi_1 \rangle \ldots \langle \Psi_n, \psi_n \rangle$ it is true that $\Psi \subseteq \Psi_1 \cup \cdots \cup \Psi_n$.

(3) The childern nodes of some node consist of all canonical defeaters of that node, that fulfil (2).

A node of a tree T is undefeated iff it has no undefeated childern. Tree T is successful iff its root is undefeated.

Now the goal of arguments is usually to establish some claim. We therefore need some criteria telling us when a certain claim can be established from a certain set of beliefs, using deductive argumentation. Naturally, we will need to take in account all argument trees for arguments that establish this conclusion, but argument trees for arguments that establish its negation as well. Authors of (Besnard & Hunter, 2001) introduce following concepts.

Definition 5 An argument structure for sentence ϕ with respect to set of sentences Δ is a pair $\langle \mathsf{P}, \mathsf{O} \rangle$, such that $\mathsf{P} = \{\,\mathsf{a} \in [\Delta] : \mathrm{Con}(\mathsf{a}) = \phi\,\}$ and $\mathsf{P} = \{\,\mathsf{a} \in [\Delta] : \mathrm{Con}(\mathsf{a}) = \neg\phi\,\}$.

To determine to what degree does a certain set of sentences support a different sentence we first construct an argument structure listing all arguments that can be made on basis of this set for this claim or against it (these sets can be empty, of course). Then we apply a certain function, called categoriser, which assigns each of these trees a certain number.

Besnard and Hunter propose several categorisers, but I will only introduce one, which they call a binary categoriser and which simply assigns to a successful tree value 1, otherwise value 0.

Further we use another function, called the accumulator, to accumulate values of all arguments both in P and in O and finally simply compare them.

Besnard and Hunter again suggest several such functors, including sum and max accumulators, but what kind of accumulator we use will be unimportant for the results I aim to demonstrate, provided we use the binary categoriser. Let us just make a convention, that each accumulator assigns 0 to empty set of argument trees.

Now let's illustrate these definitions on a practical example. Let $\Delta = \{p \vee q, q \vee p, \neg p, \neg q\}$ and $\phi = p$. How does argument structure for ϕ with respect to Δ looks like and what do we obtain by applying binary categoriser and different accumulators to it?

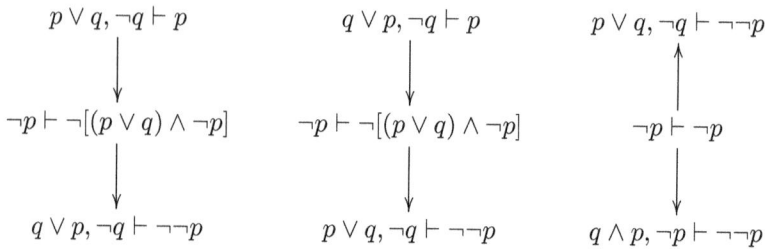

By using the binary categoriser we obtain a pair $\{\{1,1\},\{0\}\}$. Using a max accumulator we obtain a pair $\{1,0\}$, by using the sum accumulator we obtain pair $\{2,0\}$. Either way, we obtain that the given set supports the claim ϕ and does not support its negation.

Now what would happen, if we replace the condition (3) in the definition of the argument tree with a condition, allowing us to use all defeaters? We would obtain the following, much richer, argument structure:

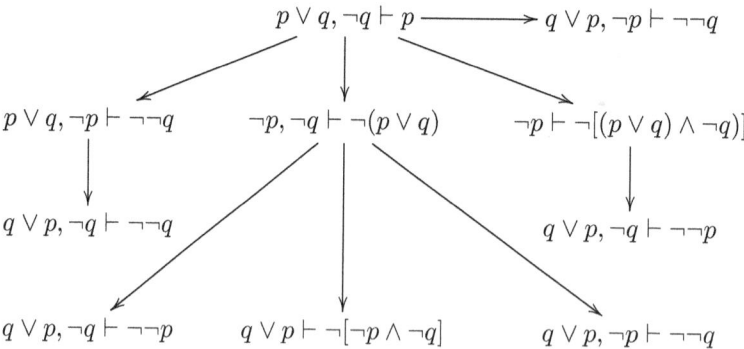

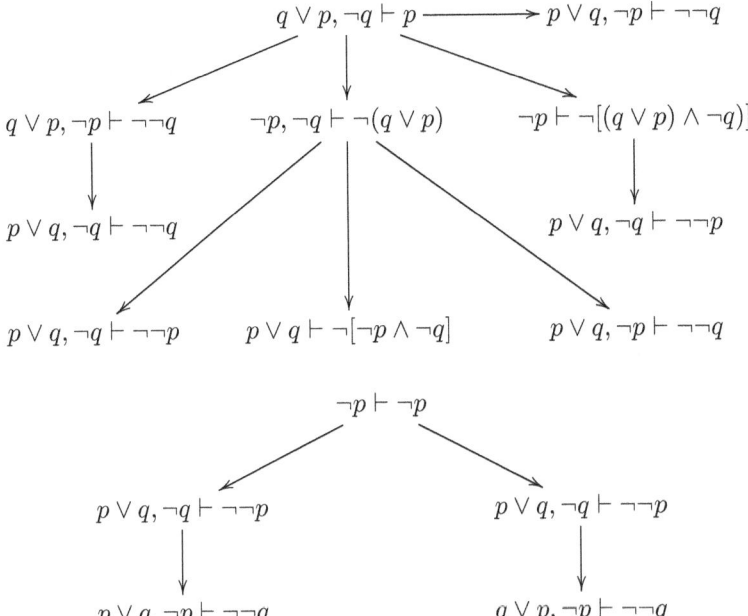

By using the binary categoriser we obtain a pair $\{\{0,0\},\{1\}\}$. Using either a max or sum accumulator we obtain a pair $\{0,1\}$. A completely opposite result from the previous case. In this case by saying we do not need other than canonical defeaters we have done injustice to the oponent of the claim that ϕ.

Notice what happened here. By allowing the opponent to argue against a stronger claim than necessary, he should have exposed himself to larger sets of possible counterarguments, but instead he actually nullified those by 'depleting' the premises, which could have been used against him, in particular the disjunction equivalent to the one used by the proponent in the argument at the root.

This is a peculiar feature of monological argumentation. Now if the proponent and the opponent would have to draw their arguments from seperate sets of beliefs, that would be a different story.

This strongly contradicts our intuition about principles governing natural argumentation. Being a more eloquent should not be a viable strategy, but it should not be simply prohibited. By simply preventing such strategy by excluding noncanonical arguments we may conceal that the condition (2) in the definition of the argument tree is rather arbitrary and unnatural and is responsible for this mismatch.

The sole purpose of this restriction, of course, is to prevent infinite branches of cycling arguments, which would necessarily occur in each argument tree, by requring a new piece of information being used at each step of the debate. Provided the set of beliefs, from which we are only allowed to draw premises of our arguments is finite, this is enough to ensure, that the argumentation will eventually come to the end.

The above examples show us that Besnard's and Hunter's model of monological deductive argumentation still requires some development. Providing new sufficient and necessary restrictions on construction of argument trees, that would prevent infinite cycling, but would not require ruling out noncanonical arguments is very interesting and challenging problem, which I have not yet solved sufficiently.

Originally, I thought that by replacing the condition (2) with a stricter one, forbidding repetition of any of the premises would make an argument undefeated in the canonical tree if and only if it was undefeated in the full tree, but recently I have discovered a flaw in the proof, so unfortunately I cannot publish it here.

3 Conclusion and future work

Deductive argumentation of Besnard and Hunter not only promises to offer new perspective on the philosophical problem of modelling human argumentation using logic and provides a tool to be used for dealing with inconsistent databases, but offers new areas of purely mathematical problems.

It might be interesting to explore variations of the original definitions and see how would strengthening or weakening of some of the requirements affect the resulting measures of credibility of some piece of information given a belief set. I have shown that only a slight modification in definition may sometimes lead to completely opposite results.

References

Bench-Capon, T. J. M., & Dunne, P. E. (2007). Argumentation in artificial intelligence. *Artificial Inteligence*, *171*, 619–641.

Besnard, P., & Hunter, A. (2001). A logic-based theory of deductive arguments. *Artificial Intelligence*, *128*, 203–235.

Bondarenko, A., Dung, P. M., Kowalski, R. A., & Toni, F. (1997). An abstract, argumentation-theoretic approach to default reasoning. *Artificial Intelligence*, *93*, 63–101.

Dung, P. M. (1995). On the acceptability of arguments and its fundamental role in nonmonotonic reasoning, logic programming and n-person games. *Artificial Intelligence*, *77*, 321–357.

Svatopluk Nevrkla
Department of Logic, Faculty of Arts, Charles University in Prague
Celetná 20, Praha 1
e-mail: svata@logici.cz

Logical Form and Reflective Equilibrium

Jaroslav Peregrin   Vladimír Svoboda*

Abstract

Though, at first sight, logical formalization of natural language sentences and arguments might look like an unproblematic enterprise, the criteria of its success are far from clear and, surprisingly, there have only been a few attempts at making them explicit. This paper provides a picture of the enterprise of logical formalization that does not conceive of it as a kind of translation from one language (a natural one) into another language (a logical one), but rather as a construction of a 'map' of (a piece of) the 'inferential landscape' of the natural language. The criteria that appear to govern the enterprise are labeled as those of reliability, ambitiousness, transparency and parsimony. These criteria, it is argued, do not provide for an excavation of a ready-made logical structure, but rather help us achieve a "reflective equilibrium" between the normative authority of logic and the answerability of logic to a natural language.

Keywords: logical analysis, logical form, reflective equilibrium, reasoning

1 Introduction

One of the most characteristic types of tasks that students of logic must deal with is usually articulated as follows: "Rewrite the following argument in logical notation and then decide whether it is valid or not". This seems quite natural—the ability to examine the correctness of argumentation is precisely what students of logic are supposed to learn. Fulfilling this kind of task consists of two parts,

---

*Work on this paper was supported by the research grant No. P401/10/1279 of the Czech Science Foundation.

each of which requires a somewhat different skill. First, students must rewrite the natural language sentence into a logical formalism (chosen by the teacher) and then they must employ a method (either already implied by the previous choice or chosen by the teacher) leading to the decision.

The first part of this enterprise is usually called *logical analysis* or *logical formalization*. Though teachers usually allow for some variations in fulfilling this task, they are generally supposed to be able to tell whether what the students provide as their solution is correct or not. In this sense, logical analysis might seem to merely be an unproblematic enterprise that might appear difficult to the students but not to the teacher, who knows the criteria of its success. Yet, if we were to ask a randomly picked teacher how she decides whether a given formalization is correct, what criteria she actually employs, she is likely to be surprised by the question. She would probably say something to the effect that anyone who masters a logical system acquires an insight that enables her to recognize the correct formalization, similarly as a good translator is able to recognize a correct translation (even without being able to explicitly articulate any general criteria).

But can we accept a response of this kind? Is formalization simply a translation from one language (a natural one) into another language (a formal one)? And is it enough to leave it on the level of an implicit know-how? We think that the answers to both these questions are negative. We do not believe that formalization is very similar to translation. And even if it were, we do not believe that it could, in general, be left at the level of practical know-how without an explicit reflection of its criteria—even translation from one natural language into another must be explicitly reflected upon once competing proposals appear.

In this respect, we find it surprising how little attention questions of this kind receive in the (meta)logical literature. In fact, the only book-length treatment explicitly devoted to the criteria of logical formalization that we know of is (Brun, 2003).[1] This is also peculiar in view of the fact that this issue is closely related to questions regarding the very nature of logic, especially the question to what extent logic is merely descriptive and to what extent is it prescriptive.

---

[1] Many not so systematic considerations of this kind are, of course, scattered throughout the literature, especially about 'logical form', ranging from the seminal paper of Russell (1905) to more contemporary treatises like Sainsbury (1991).

In this paper, we want to contribute to the self-understanding of logical analysis of natural language by examining the practice of formalizing natural language sentences and arguments more explicitly than is common among the practitioners of logical analysis. We will try to articulate the criteria of correctness implicit to this practice and we will also try to draw some conclusions regarding the nature of logic. Some of the conclusions that we reach may appear as sheer platitudes, others as quite controversial. We believe that even the platitudinous parts are something that must be stated explicitly, and we believe that the controversial parts can be justified. This is what we will attempt to show in the rest of the paper.

2 Translation or not?

The usual idea is that as formalization, *viz.* rewriting a given sentence or argument into a logical language, is a kind of translation, it can be assessed like any other translation—the most basic criterion of the success of the enterprise being the preservation of meaning. What we want to point out in this section is that this is misguided. Though formalization may involve a translation-like step, it should not, in general, be considered as translation. We must therefore look for different criteria of its success than those governing translation.[2]

A student who faces the task of formalizing a sentence of natural language usually starts from paraphrasing the sentence in such a way that the resulting sentence (still in the natural language) reflects the shape of a suitable logical formula, and then proceeds to an expression of a language that mixes expressions of natural language with logical symbols. We will call expressions mixing natural language phrases in their raw form with phrases of a logical language *hybrid expressions*.

---

[2]Let us note that the problem of adequate formalization of a sentence formulated in a natural language can be considered from two different perspectives: a narrower one and a wider one. The narrower, *internal* one consists in accepting a framework of a particular logical language; the wider, *external* one does not presuppose such a framework and considers the choice of the language as a part of the task to be solved. (And we should keep in mind that we need not even restrict ourselves to the spectrum of existing languages—we could even consider the possibility of inventing a new one.) In this article we will, most of the time, be adopting the internal perspective. We will operate within the framework of classical predicate logic (CPL).

The process in which we 'translate' a natural language sentence into such a formula can be called, following Quine, *regimentation*.

Let us, for example, consider the following simple sentence

>   (S1) *Dogs have legs.*

A student who is to formalize it in classical predicate logic (CPL) (and is already sufficiently indoctrinated) is likely to see it as 'in fact' a shortcut for

>   (S1′) *All dogs have legs*

and to see this sentence as saying what is more precisely expressed by:

>   (S1″) *For every individual it is the case that if it is a dog, then it has legs.*

So far, what has been going on is paraphrasing *within English*, but now it seems natural (to logicians) to take a further step and make the meaning of the last sentence even more precise and transparent by means of CPL. Thus, we get the following hybrid expression

>   (HF1) $\forall x(\textbf{\textit{Dog}}(x) \rightarrow \textbf{\textit{Has-legs}}(x))$

where $\forall$ and $\rightarrow$ are the constants whose meanings (or 'meanings') are exactly delimited, whereas ***Dog*** and ***Has-legs*** are terms about which we merely presuppose that they inherit the meaning of the English expressions *is a dog* and *has legs*.

Thus we achieve a logical regimentation.[3] Note that its outcome—a hybrid formula—is not an expression of a language with a coherent semantics, for it consists of two different kinds of constituents, the respective semantics of which are of very different natures. Due to the fact that it contains elements of natural language (with their natural meanings) it is not a formula of a logical calculus, while due to the

---

[3]Of course, that regimentation may involve significant shifts in meaning. While (HF1) is false once there is a single individual dog which is—perhaps by some strange coincidence—legless, hardly anybody would consider (S1) as false in such a situation. Moreover, regimenting (S1) as (HF1) involves another meaning shift: While normal speakers would probably not hesitate to infer that there are some dogs that have legs from (S1), it is not correct to infer $\exists x(\textbf{\textit{Dog}}(x) \wedge \textbf{\textit{Legs}}(x))$ from (HF1).

fact that it contains artificially introduced symbols it does not have (strictly speaking) a natural meaning. Nevertheless, people conversant with the corresponding logical system can understand them very well—or at least this is what they feel.

There are two obvious pathways leading from such a hybrid language to a logical language whose semantics is fully and explicitly delimited (i.e. its expressions are put together according to explicit formation rules and their functions or semantic values are explicitly given). The first one consists in also rectifying the 'extralogical' vocabulary of natural language, thus gaining, aside from *logical* constants, *extralogical* ones as well and, as a consequence, formulas that contain no elements of the vocabulary of natural language.[4] (In the case of extralogical vocabulary it is, however, much less clear how to capture its functioning.) In this way, we reach a language which is *formalized*, though not *formal*—in the sense of Tarski (1933).[5]

Taking this kind of step in our example yields the formula

(CF1) $\forall x(\boldsymbol{D}(x) \rightarrow \boldsymbol{L}(x))$

What are $\boldsymbol{D}$ and $\boldsymbol{L}$ here? As they must be expressions belonging to a language within the framework of CPL and hence with logical (mathematized) semantics, there seems to be only one option—they are unary predicate letters, and hence they denote subsets of the universe. Of course, if the bold letters are given this meaning, (CF1) will be true or false (unchangeably) depending on the relations of the particular subsets they represent.

The second pathway consists in *dismissing* the extralogical constants. This is a natural thing to do if what we are after is a fully-fledged *formalization* that leads to a language that is not just formalized, but *formal*, in the sense that its formulas do not correspond to natural language sentences but rather to sentence *forms*. The step from an expression like (HF1) to a formal language expression consists in dropping the terms borrowed from natural language and replacing them with utterly meaningless symbols, which we will call *parameters*. Let us call this step away from the hybrid language *abstraction* (as

---

[4] The boundary between the 'logical' and 'extralogical' vocabulary of natural language is, of course, blurry.

[5] A paradigmatically clear case of this is Peano arithmetic: its language consists, aside from the logical constants, of the extralogical constants $\boldsymbol{0}$, $\boldsymbol{S}$, $\boldsymbol{+}$, and $\boldsymbol{\cdot}$, whose functioning is exactly stipulated.

we abstract from meanings of certain expressions). In our case, we obtain the traditional formalization

(FF1) $\forall x(F(x) \to G(x))$

Let us stress, once again, that (FF1) is no longer a meaningful sentence, but rather a *pure* formula, i.e. an articulation of a mere sentence *form*, containing meaningless parameters (we will speak simply about a *formula*, where no confusion is likely). It is sometimes also referred to as the *logical form* of the sentence out of the regimentation of which it has been abstracted. The hybrid formula that served as the input of the abstraction may be called an *instance* of the form; other instances are all those hybrid formulas that result from the replacement of the parameters of the formula by natural language terms of suitable grammatical categories. For simplicity, we will call the natural language sentences that verbalize the instances of a formula its *natural language instances*.

Thus, for us, the outcome of the formalization of a sentence like (S1) is a (pure, i.e. 'uninterpreted') formula such as (FF1). However, logicians engaged in logical analysis of language sometimes do not see it in this way: what they consider as the result of formalization is rather a ('fully interpreted', possibly hybrid) formula of the kind of (CF1) or (HF1). This might be an innocent terminological clash solvable simply by acknowledging the ambiguity of the term *formalization*, choosing one of the senses and introducing a different word for the other. Curiously enough, however, some of the most prominent logicians engaged in the logical analysis of language try to avoid this choice. They propose something that appears to be a somewhat strange compromise between the two options. This approach, promoted, among others, by Sainsbury (1991); Brun (2003) or Baumgartner and Lampert (2008), would result in the articulation of the result of regimentation of (S1) in the following shape:

(FC1) $\forall x(F(x) \to G(x))$
     F: ... is a dog; G: ... has legs

Here the first line, which is nothing other than (FF1), is complemented by the second one, which is called the *correspondence scheme*.[6] Thus,

---

[6] Sainsbury (1991) and Brun (2003) use the term "correspondence scheme", Baumgartner and Lampert (2008) call the same thing "realization".

the outcome of the formalization of a natural language sentence is (somewhat surprisingly) not a formula of a logical language but something more complex. What is the nature of such complexes? One possibility of how to read them is to take the correspondence scheme as simply an instruction for the interpretation of the parameters; hence, in our case, as an instruction to interpret F by a certain set (the extension of *is a dog* at our world at some time point) and G as another one (that of *has legs*). In this case, (FC1) would simply collapse into (CF1), and there would seem to be no reason for presenting the result in this complex form.

Another option is to read (FC1) as establishing a 'dynamic' connection between the parameters and their natural language counterparts—in the sense that the latter confer their extensions on the former not on a one-time basis, but continually, which results in a situation where the extensions of F and G repetitively change. In this way we can say that (FC1) has not only a constant truth *value*, but truth *conditions*—it has (similarly as (S1)) different truth values in different situations/possible worlds. It is, however, surprising that formalization of a sentence in CPL does not yield a formula with the standard semantics but one with a kind of 'intensional' semantics. We are afraid that this institutes a dangerous Janus-facedness of (FC1): on the one hand, it is seen as a formula of (CPL) (disregarding, in effect, the second line), while on the other it is seen as an 'intensional' formula with non-trivial truth-conditions.[7]

We think this is a mere trick: a trick that supplies a first-order formula with truth *conditions* when, in fact, it has merely an (unchangeable) truth *value*. Moreover, we think it is an *unnecessary* trick: the criteria of adequacy of a logical formalization need not be (and, in fact, are not) based on the comparison of truth conditions, but rather on the comparison of behavior within arguments. Recognizing the logical form of a sentence is, first and foremost, recognizing the correctness/incorrectness of the arguments in which the sentence features, i.e. identifying its inferential role. Doing logical formalization, we start from a natural language argument, move to its logical form in a logical system and then use the means of the logical system to decide whether the argument form is logically valid—where the move from natural language to the formal one is usually not di-

---

[7]See Peregrin and Svoboda (in press) for a more detailed discussion.

rect but leads via the intermediate level of a hybrid language. And while the first part of the move (from natural to the hybrid language) might perhaps be considered as a kind of translation, the second part (from the hybrid to the formal language) certainly does not have this character.

3   How can we assess adequacy of logical formalization?

Suppose that three students are given the task to formalize the sentence

(S2) *No red snakes are dangerous*

and they came up with the following respective proposals:

(FFS2a) $\neg\exists x((Fx \wedge Gx) \rightarrow Hx)$

(FFS2b) $\neg\exists x(Fx \wedge Gx \wedge Hx)$

(FFS2c) $\forall x((\neg Gx \vee \neg Fx) \rightarrow \neg Hx)$

(where the parameter F replaces the expression *is red*, G replaces *is a snake*, and H replaces *is dangerous*). How could we find out which of the proposals is to be preferred?

Unlike those who propose the 'corresponding schemes' as part of the result of formalization, we cannot take recourse to the sameness of truth conditions—the above formulas, not being sentences of a fully-interpreted language, simply do not have any. But we have already indicated what we should focus on instead: the behavior in arguments, i.e., in effect, inferential roles. What we usually do, as a matter of fact, is a careful reflection on arguments of a certain kind. We can consider (implicitly or explicitly) a sample list of natural language 'reference arguments' that we intuitively hold for falling into the intended scope of the logical system we use (here CPL)[8] and that are *perspicuous* in

---

[8] We assume that each logical system has been conceived with the goal of accounting for the behavior of a certain part of the logical vocabulary of natural language and the arguments that hold in virtue of this very vocabulary. Classical propositional logic focuses on the behavior of the well known connectives, classical predicate logic adds the basic quantifiers to this and modal logic further adds a certain modal vocabulary, etc. The intended scope of the system is then constituted by the arguments that are correct solely in virtue of the specific kind of vocabulary that the logical system is supposed to capture.

Logical Form and Reflective Equilibrium

the sense that each of them is clearly intuitively correct or incorrect and in which the sentence we are considering (here S2) features as a premise or as the conclusion. Let us call the arguments on such a list *reference arguments of the sentence*. In our case, for example, a list of reference arguments can contain the following (correct and incorrect) cases:

> *Kaa is red*
> *Kaa is a snake*
> *Kaa is not dangerous*
> ―――――――――――――――――――
> *No red snakes are dangerous*

> *Every snake is a reptile*
> *No reptile is dangerous*
> ―――――――――――――――――――
> *No red snakes are dangerous*

> *No red snakes are dangerous*
> *Kaa is not red*
> ―――――――――――――――――
> *Kaa is not dangerous*

> *No red snakes are dangerous*
> *Kaa is not dangerous*
> ―――――――――――――――――
> *Kaa is red*

> *No red snakes are dangerous*
> *Kaa is a red snake*
> ―――――――――――――――――
> *Kaa is not dangerous*

If we now, next to the arguments, put parallel lists consisting of argument forms composed of the corresponding formulas of CPL, in which the sentence *No red snakes are dangerous* is formalized in each of the three proposed ways respectively, we get the table printed on the next page. For a better orientation we write those sample arguments that are (intuitively) correct with bold font and similarly for the argument forms that are valid in CPL.[9]

How does this list help us decide which of the proposed formalizations of (S1) is the most adequate one? The general answer is obvious: Where we have an intuitively incorrect argument that is rendered as valid by its formalization, or where we have, conversely, an intuitively

―――――――――――
[9] F, G, H are as before, l replaces *is a reptile* and k replaces the name Kaa.

Kaa is red Fk
Kaa is a snake Gk
Kaa is not dangerous Hk
―――――――――――――――――――   ―――――――――――――――
No red snakes are dangerous $\neg\exists x(Fx \land Gx \land Hx)$

Fk
Gk
Hk
―――――――――――――――
$\neg\exists x((Fx \land Gx) \to Hx)$

Fk
Gk
Hk
―――――――――――――――
$\forall x((\neg Gx \lor \neg Fx) \to \neg Hx)$

Every snake is a reptile $\forall x(Gx \to Ix)$
No reptile is dangerous $\neg\exists x(Ix \land Hx)$
―――――――――――――――――――   ―――――――――――――――――
No red snakes are dangerous $\neg\exists x((Fx \land Gx) \to Hx)$

$\forall x(Gx \to Ix)$
$\neg\exists x(Ix \land Hx)$
―――――――――――――――
$\neg\exists x(Fx \land Gx \land Hx)$

$\forall x(Gx \to Ix)$
$\neg\exists x(Ix \land Hx)$
―――――――――――――――
$\forall x((\neg Gx \lor \neg Fx) \to \neg Hx)$

No red snakes are dangerous $\neg\exists x((Fx \land Gx) \land Hx)$
Kaa is not red $\neg Fk$
―――――――――――――――――――   ―――――――
Kaa is not dangerous $\neg Hk$

$\neg\exists x(Fx \land Gx \land Hx)$
$\neg Fk$
―――――
$\neg Hk$

$\forall x((\neg Gx \lor \neg Fx) \to \neg Hx)$
$\neg Fk$
―――――
$\neg Hk$

No red snakes are dangerous $\neg\exists x((Fx \land Gx) \to Hx)$
Kaa is not dangerous $\neg Hk$
―――――――――――――――――――   ―――――
Kaa is not red $\neg Fk$

$\neg\exists x(Fx \land Gx \land Hx)$
$\neg Hk$
―――――
$\neg Fk$

$\forall x((\neg Gx \lor \neg Fx) \to \neg Hx)$
$\neg Hk$
―――――
$\neg Fk$

No red snakes are dangerous $\neg\exists x(Fx \land Gx \land Hx)$
Kaa is a red snake $Fk \land Gk$
―――――――――――――――――――――   ―――――――
Kaa is not dangerous $\neg Hk$

$\neg\exists x((Fx \land Gx) \to Hx)$
$Fk \land Gk$
―――――
$\neg Hk$

$\forall x((\neg Gx \lor \neg Fx) \to \neg Hx)$
$Fk \land Gk$
―――――
$\neg Hk$

correct argument that is rendered as incorrect, the formalization becomes suspicious. Thus, the fourth and the fifth case suggest that we have a reason to reject the formalization (FFS2a), whereas the second and the third cases provide reasons for rejecting (FFS2c). Hence the victorious formalization that we (tentatively) embrace is (FFS2b), which was not 'disproved' by the reference arguments.

Let us note that this method of selecting the best formalization is, in fact, not so different from that employed by the adherents of 'correspondence schemes'. The point is that inspecting the correctness of arguments, such as that which we have just been engaged in, can be seen as inspecting truth conditions. For example, the claim that the second argument is correct can be read as the claim that (S2) is true in all situations where all snakes are reptiles and no reptile is dangerous. If we consider formulas (FFS2a), (FFS2b) and (FFS2c) furnished by the 'correspondence schemes', we can say that the first two of them are also true in all such situations (where the situations are described in terms of the corresponding language). The same thing, however, cannot be said about (FFS2c). (Inspecting other arguments may directly amount to inspecting the truth conditions of sentences other than (S2), with (S2) taking part in the characterization of the situations considered).

In general, what we actually do when we check for the truth value of a sentence in a certain situation is, in fact, hardly distinguishable from checking inferences. We must somehow characterize the situation which we are considering, and we can hardly do it otherwise than in terms of some sentences; hence, when we then ask whether a sentence is true in the situation, we can be seen as asking whether the latter sentence follows from the former ones. (It is true that the sentence in which we characterize the situation can be couched in a metalanguage rather than in the object language we are analyzing; but, if it is natural language that is our ultimate target, then we cannot count on a metalanguage different from it.)

4 Criteria

How to articulate criteria of adequacy of formalization based on the above insights? If we generalize the lesson from the sketch of the method presented in the previous section, we can say that the point of the formalization is to make explicit the place of a natural language sentence A within the inferential structure of its natural language, by means of associating A with a formula of the logical system **S** the position of which within the inferential structure of **S** is explicit and definite. Hence, with the help of **S** we construct a 'map' of the 'inferential surroundings' of A, making it possible for us to gain an overview over this 'inferential landscape'. This allows us to spot the inferential interrelationships of A with other sentences, which would be not so easily discernible otherwise.

However, it is crucial to keep in mind that if we try to identify the inferential (sub)structures of a natural language we want to make explicit, we will necessarily uncover a slightly fuzzy and gappy network of relations among sets (or sequences) of sentences (premises) and individual sentences (conclusions). The inferential structure of **S** will be, on the other hand, definite, determinate and much simpler.

To be able to formulate the criteria of adequacy of logical formalization that has issued from the above considerations, we introduce some terminology. A $[\Phi/A]$-*formalization* of an argument containing A will be a formalization with the formula Φ in place of A; conversely, a $[\Phi/A]$-*instance* of an argument form containing Φ will be any natural language instance of the form with A in place of Φ. Thus, given that A is *All dogs have legs* and Φ is $\forall x(P(x) \to Q(x))$, the $[\Phi/A]$-*formalization* of the argument

$$(A1) \frac{\begin{array}{c} \text{All dogs have legs} \\ \text{Fido is a dog} \end{array}}{\text{Fido has legs}}$$

will be (given that the formalizations of *Fido is a dog* and *Fido has legs* are fixed as $P(a)$ resp. $Q(a)$):

$$(AF1) \frac{\begin{array}{c} \forall x(P(x) \to Q(x)) \\ P(a) \end{array}}{Q(a)}$$

Conversely, (A1) will be an $[\Phi/A]$-*instance* of (AF1).

Now an argument form containing Φ is $[\Phi/A]$-*defeated* if it has an intuitively incorrect $[\Phi/A]$-*instance* among the reference arguments representing the intended scope of the actual logic (otherwise it is

Logical Form and Reflective Equilibrium

[Φ/A]-*undefeated*). Given this terminology, we can articulate the most fundamental criterion of adequacy of formalization, which we will call the *principle of reliability*, rather succinctly:

(REL) Φ is a *proto-adequate formalization* of A in **S** iff no argument form valid in **S** and containing Φ is [Φ/A]-defeated.

The other criterion implicit to our proceedings envisaged in the previous section can be termed the *principle of ambitiousness*:

(AMB) Among the proto-adequate formalizations of A, Φ is the more adequate formalization of A in **S** the more intuitively correct arguments belonging to the intended scope of **S** in which A features as a premise or a conclusion are rendered as valid argument forms of **S**.[10]

To complete a truly comprehensive set of criteria we should add some principles guiding the choice for the cases undecided by the previous criteria. They can be called *the principle of transparency* and *the principle of parsimony*. We can articulate the first principle, for example, in this way:

(PT) (Other things being equal,) Φ is the more preferable formalization of the sentence A in the logical system **S** the more the grammatical structure of Φ is similar to that of A.

The second principle can then be formulated as follows:

(PP) (Other things being equal,) Φ is the more preferable formalization of the sentence A in the logical system **S** the more it is parsimonious as concerns the number of (types as well as tokens) of logical symbols it employs.

The import of the principles should be seen as decreasing in the order in which they have been presented. The first of them is close to

---

[10]These two principles are similar to (COR) and (COM) of Brun (2003). We have chosen different labels as we do not want to suggest that the first of them must be inevitably fulfilled for a formalization to count as *correct* in the ordinary sense of the word (valid argument forms to which we have natural language counterexamples might be a price we are willing to pay for having a particularly simple and perspicuous logical system); and that the second one marks a *completion* which we must achieve to be successful—we rather think that it spells out an ideal which we usually want merely to more or less approximate.

a *sine qua non* matter (though keep in mind that this holds only in the realm of the intended scope of the logic in question). The second is essential as well, as it suggests that the logician should not search just for 'the safest' formalization but also for the inferentially most 'fruitful' one—the one that makes explicit more relevant valid inferences than competing ones. The last two principles are more-or-less auxiliary (though they can be given more weight within analyses made for certain specific purposes). Thus, especially in the case of the last three, there might be various trade-offs (we might, for example, want to have a regimentation that is not quite transparent if it is exceptionally parsimonious.).

5 Bootstrapping

Now, however, we must return to various simplifying assumptions that we made throughout the course of our way from the description of the praxis of logical analysis to our articulation of the criteria.

First, the *principle of correctness* states that we can consider Φ as a candidate for the formalization of A only if *no* argument form containing Φ is $[\Phi/A]$-defeated. In fact, this is not quite realistic. Sometimes we may encounter what look to be invalid instances of argument forms that we hold for valid without putting their validity into doubt. Thus, consider the following argument, which looks, at least *prima facie*, as an instance of (AF1):

(A2) $\dfrac{\textit{All dogs have common genes}\quad\textit{Fido is a dog}}{\textit{Fido has a common gene}}$

This is clearly not a valid argument. Yet, its existence is not likely to make us conclude that (AF1) is defeated by (A2)—we would rather conclude that (A2) is, despite appearances, *not* an instance of (AF1), in particular that the logical form of *All dogs have common genes* is *not* $\forall x(P(x) \to Q(x))$. Why? We will probably say something to the effect that the predicate *to have common genes* is not an 'individual-level' (but rather 'group-level') predicate and thus should be represented, on the level of logical form, in a way different from the individual-level ones. However, how do we tell such an individual-level predicate from a group-level one? We might well say that a predicate is individual-level if its use in the place of Q in instances of (AF1) yields correct arguments. But we would then have a vicious circle: an argument form is valid *because* all its instances are correct, but to be an instance of a valid form appears to *involve* being correct. (Of course the circle need not be so straightforward—we need not take directly (AF1) as the hallmark of individual-levelness of the predicate involved. However, we think that in the end *some* kind of circle is inevitable, for the distinction between individual-level and group-level predicates is not syntactic in the sense that it would be discernible by studying the predicates aside of their inferential properties.)

Is this circle vicious? Not necessarily. We think that it only points out that what we see as valid forms is not something which we can directly read off natural language, but rather that it is something that must be bootstrapped into existence. It is okay to explain away *some* invalid *prima facie* instances of an allegedly valid schema provided they can be plausibly taken as something marginal; however, if there is no way of moving them into a marginal position, we must retract the validity of the form.

Similar kinds of bootstrapping, in our view, penetrate the whole enterprise of logical formalization. Thus, we have to return to another unrealistic assumption that we have tacitly made when we started to look for the criteria of adequacy of formalization, *viz.* the assumption that the formalizations of all other sentences, save the one whose formalization we are pondering, are fixed.[11] Taken literally, it would, of course, once again lead us into a vicious circle: if we had to base the regimentation of any sentence on already accomplished formalizations of other sentences, the whole enterprise would never really be able to get out of the ground.

And once again the solution is, of course, a bootstrapping: we start with mere tentative regimentations of some simple sentences, basing the regimentations of others on them. Hence, if we are considering Φ as a possible formalization of A and we find out that some argument form involving Φ as a counterpart of A is valid, whereas there is a natural language instance that provides a counterexample (defeats the argument form), we will not only consider dropping the hypothesis that Φ is an adequate formalization of A, but will also take into

---

[11]In our case the trick was not so obvious as we only employed, in our test examples, formalizations of simple sentences that seem quite straightforward, like *Kaa is (not) a snake*.

account the possibility of keeping the hypothesis at the cost of dispensing with formalizations of some of the other sentences involved in the counterexample. Again, the process of formalization of sentences and arguments is, in fact, a holistic, give-and-take enterprise.

The third simplifying assumption was implicit to assuming our internal perspective, i.e. assuming that the logical language we use for the formalization is fixed. A formal language used as the tool of formalization is always more or less Procrustean, and to a certain extent this may be seen as its *virtue*: it lets us get rid of those elements of natural language that are irrelevant from the viewpoint of argumentation or of semantics and lets us clearly see the relevant backbone. But it might well happen that it may come to be Procrustean to the extent that it becomes a *vice*: it makes us neglect or obscure some important feature of natural language. In such a case, we need to ascend to the external perspective and look for a more suitable language.[12]

Hence, even the language we use for the formalization must be bootstrapped into existence: to a certain extent the features of natural language that do not fit into the mould of such language, are tolerable if they can be explained away as irrelevant or marginal. Once this discrepancy becomes excessive, however, it may be wise to give up on the language and upgrade. (The fact is, the standard logical languages, like those of classical propositional and predicate logic, have come to be taken so much for granted that we often take their adequacy as self-evident and tend to ignore discrepancies between them and natural language.)

6 Reflective equilibrium

The considerations of the previous section indicate that logic, though in a sense dealing with inferential patterns extracted from natural language (and thus answerable to how the language, in fact, works), also has a normative role to play: once it acquires a definite shape, it assumes the role of a standard which can be used to adjudicate individual cases of argumentation not only within a hybrid language but also in the natural one. As long as logical rules are in force, they decide what a correct argument is. But once a logical system urges us

---

[12]The common logical languages are, of course, common exactly for the reason that they turned out to be *tolerably* Procrustean.

to correct intuitions of competent speakers too frequently or in such a way that that we perceive the corrections as too counterintuitive, we have a serious reason to amend some rules of the logical system or to abandon the system as a whole. Hence, we have here the most basic give-and-take. And this is where, we believe, we must see it as a matter of what Goodman (1955) aptly called the *reflective equilibrium*.

We can, in general, say that the laws articulated by logic are not merely a reflection of something that exists, in a wholly articulated shape, either within our thinking or somewhere under the surface of our language. There is no way of merely extracting already completed laws of logic directly from there—what we can get as the starting point of logic are certain patterns of valid inferences that are accepted across different domains of our discourse and reasoning but which are not quite definite (both in the sense of not being exceptionless, and in the sense of not having an utterly clear-cut semantics).

This implies that any kind of logical system may only partially be based on patterns which logicians simply *find* and *report*—it must *also* be based on *completions* and *streamlinings* that logicians perform. Hence the laws of logic, as articulated by logicians, though crucially reflecting pre-existing patterns of valid inference, go well beyond them. Thanks to this and also to the—modest but extant—feedback that the work of logicians receives, logic influences the language of science and consequently even—slightly—the colloquial idiom, and comes to be taken as a *norm*. It acts as a norm of what is to be seen as regular and what is to be seen as 'irregular', and what is a lawful usage and what is an exception. (In this way, it ties together a framework for adjudicating various disputes that would hardly be resolvable otherwise.)

We have tried to portray how this works in terms of the dialectics of correct inferences and valid forms. Some inferences (in natural language) are *prima facie correct*, which makes us see some forms of inferences (namely those which have correct instances) as *prima facie valid*. However, we take the quest for (getting a grasp on) validity as an instance of a quest for *e pluribus unum*, as a quest for finding a perspicuous order within the *prima facie* messy vastness of individual cases of more or less correct or incorrect inferences; this makes us impose more order on our language and our reasoning than we are able to *find* there, even at the cost of some Procrustean trimming and stretching. Hence, upon reflection, a form of inference comes to be taken as valid not exactly in those cases when all its natural language instances are correct, but in cases when those which are not can be reasonably explained away.

More traditional approaches to logical formalization often create the illusion that, behind or beneath the surface form of our language, there is a definite deeper and more substantial logical form. However, we do not believe that anybody could get to such a form by a process substantially different than the 'give-and-take' one described above, hence by a process led by the maxim of simplicity and maximal order—the maxim that is operative in any science. In particular, we do not believe that we can get from the surface form to the logical form by some process that has nothing to do with the considerations described above under the heading of *reflective equilibrium* and we don't therefore think that logic could be left with the task of pulling out the ready-made structure and lending it a perceptible form. We are convinced that the way from the surface to the so-called logical form involves considerations largely constitutive of logic, so that the resulting logical form is not what logic merely describes or reports, but rather what logic helps bring into being.

According to this picture, logical formalisms basically generalize and systematize the inferential and semantic features of natural language and so they are liable to criticism as other empirical generalizations. However, due to the fact that natural language is vague and open-ended, formalization also does the job of sharpening, explicating and removing inconsistencies; and, as a consequence of this, the result gains a certain normative authority over the use of means of natural language.

7 Conclusion

Accepting a certain logical system, we typically proceed by *regimenting* a natural language sentence into a hybrid sentence/formula, from which we then *abstract away* the (extralogical) remnants of natural language thus reaching formulas that represent what is traditionally called the *logical form* of the sentence (in the language of the given system of logic). The most basic of the criteria governing this enterprise can be termed the criterion of reliability; it is supplemented by the criteria of ambitiousness, transparency and parsimony. The criteria do not guarantee that there is anything like a unique logic form to be found. Especially the latter three operate on a give-and-take basis, but even the first is not essential in the sense that it would be absolutely non-negotiable.

Logic aims at bringing order to our argumentative practices, by means of achieving the *reflective equilibrium*. Thus, logic has a certain descriptive aspect in the sense that it has to reflect the basic inferential structures of natural language, but it also has a normative aspect in the sense that once established, it has a (limited) authorization to brand natural language arguments as correct or incorrect.

References

Baumgartner, M., & Lampert, T. (2008). Adequate formalization. *Synthèse, 164*, 93–115.

Brun, G. (2003). *Die richtige Formel*. Frankfurt: Ontos.

Goodman, N. (1955). *Fact, fiction, and forecast*. Cambridge (Mass.): Harvard University Press.

Peregrin, J. (2008). What is *the* logic of inference? *Studia Logica, 88*, 263–294.

Peregrin, J. (2010). The myth of semantic structure. In P. Stalmaszczyk (Ed.), *Philosophy of language and linguistics, vol. I: The formal turn* (pp. 183–197). Frankfurt: Ontos.

Peregrin, J., & Svoboda, V. (in press). Criteria for logical formalization. *Synthèse*. (already available online)

Quine, W. (1960). *Word and object*. Cambridge (Mass.): MIT Press.

Russell, B. (1905). On denoting. *Mind, 14*, 479–493.

Sainsbury, R. (1991). *Logical forms (an introduction to philosophical logic)*. Oxford: Blackwell.

Tarski, A. (1933). *Pojęcie prawdy v językach nauk dedukcyjnych*. (English translation The concept of truth in formalized languages. In Tarski, 1956, *Logic, semantics, metamathematics* (pp. 152–278), Oxford: Clarendon Press.)

Jaroslav Peregrin and Vladimír Svoboda
Department of Logic, Institute of Philosophy
Academy of Sciences of the Czech Republic
Jilská 1, 110 00 Praha 1
Czech Republic
e-mail: jarda@peregrin.cz, svoboda@site.cas.cz

Implications as Rules in Dialogical Semantics

Thomas Piecha   Peter Schroeder-Heister*

Abstract

The conception of implications as rules is interpreted in Lorenzen-style dialogical semantics. Implications-as-rules are given attack and defense principles, which are asymmetric between proponent and opponent. Whereas on the proponent's side, these principles have the usual form, on the opponent's side implications function as database entries that can be *used* by the proponent to defend assertions independent of their logical form. The resulting system, which also comprises a principle of cut, is equivalent to the sequent-style system for implications-as-rules. It is argued that the asymmetries arising in the dialogical setting are not deficiencies but reflect the pre-logical ('structural') character of the notion of rule.

Keywords: dialogues, rules, sequent calculus, proof-theoretic semantics, cut

1 Introduction

Various constructive interpretations of implication have been proposed, the most prominent being those based on or related to the Brouwer–Heyting–Kolmogorov (BHK) interpretation[1]. The latter are based on the *transmission view*, according to which a proof of an implication $A \to B$ consists of a constructive procedure which transforms

---

*This work has been supported by the ESF research project "Dialogical Foundations of Semantics (DiFoS)" within the ESF-EUROCORES programme "LogICCC – Modelling Intelligent Interaction" (DFG Schr 275/15-1) and by the French-German ANR-DFG project "Hypothetical Reasoning" (DFG Schr 275/16-1/2).

[1] Cf. (Heyting, 1971; de Campos Sanz & Piecha, 2011).

any given proof of A into a proof of B. The dialogical or game-theoretical interpretation in Lorenzen-style dialogues[2] can be viewed as a variant of it: An implication $A \to B$ is attacked by claiming A and defended by claiming B. This means that in order to have a winning strategy for $A \to B$, the proponent must be able to generate an argument for B depending on what the opponent can offer in defense of A. In contradistinction to standard constructive interpretations, the attacker need not necessarily spell out a full proof of A. Instead, the proponent may force the opponent to produce certain fragments of a proof of A that are sufficient to successfully defend B. In this sense one may speak of a *partial* or *piecemeal* transmission view as being present in this approach.

2 Implications as rules

There is a more elementary view of implication, which is not based on transmission, but on the view of $A \to B$ being a rule, which allows one to pass over from A to B. This view is particularly supported by the treatment of implication in natural deduction. There modus ponens can be read as the application of $A \to B$ as a rule, which is used to pass from A to B, that is, modus ponens can be read as a schema of rule application. The introduction of an implication $A \to B$ can be read as establishing a rule, namely by deriving its conclusion B from its premiss A. Applications of logic such as logic programming or deductive databases support this perspective. Reading implications as rules motivates an alternative implication-left schema

$$(\to \vdash)^\circ \; \frac{\Gamma \vdash A}{\Gamma, A \to B \vdash B}$$

in Gentzen's sequent calculus for intuitionistic logic, yielding what we call the sequent calculus LI°. This schema expresses that by assuming the implication-as-rule $A \to B$ we are entitled to infer B from A. When reading implications as rules, we give implication an elementary meaning which is conceptually prior to the meaning of other operators. In particular, it is explained independent of harmony or symmetry considerations that would normally apply to logical connectives, simply because it is more elementary.

---

[2] See e.g. (Lorenzen, 1960; Sørensen & Urzyczyn, 2006, Ch. 7; Felscher, 1985, 2002).

Implications as Rules in Dialogical Semantics 213

The relationship to Gentzen's standard schema is spelled out in (Schroeder-Heister, 2010). Here we just point out that LI° does not have the cut elimination property. The sequent $a, a \to (b \wedge c) \vdash b$ (for atomic and distinct formulas a, b, c) can only be derived by using (Cut):

$$(\to \vdash)^\circ \; \frac{\text{(Id)} \; \overline{a \vdash a}}{a, a \to (b \wedge c) \vdash b \wedge c} \quad (\wedge \vdash) \; \frac{\text{(Id)} \; \overline{b \vdash b}}{b \wedge c \vdash b} \qquad (1)$$
$$(\text{Cut}) \; \frac{}{a, a \to (b \wedge c) \vdash b}$$

This is the only kind of derivation where (Cut) cannot be eliminated.

Although LI° does not have the cut elimination property, it does have the *weak cut elimination property*. That is, every LI°-derivation containing an application of (Cut) can be transformed into an LI°-derivation of the form

$$(\to \vdash)^\circ \; \frac{\vdots}{\Gamma \vdash A} \quad \frac{\vdots}{\Delta, A \vdash C} \qquad (2)$$
$$(\text{Cut}) \; \frac{}{\Gamma, \Delta \vdash C}$$

where the left premiss of (Cut) is the conclusion of an application of $(\to \vdash)^\circ$. Furthermore, the right premiss of (Cut) can be assumed to be either the conclusion of a derivation of the above form, or it is the endsequent in a derivation such that the cut formula A is the result of an application of a left introduction rule in the last step. As a consequence of the weak cut elimination property, LI° has the subformula property.[3]

3 Dialogical semantics

In what follows, we carry the implications-as-rules approach over to the framework of dialogical semantics. Once an implication $A \to B$ has been claimed by the opponent, it is considered to be a rule in a sort of 'database', which later on can be used by the proponent in order to reduce the justification of its conclusion B to that of A. This is achieved by allowing the proponent to defend an attack on B by asserting A whenever $A \to B$ has been claimed by the opponent before. In case no such claim has been made before (i.e., if no applicable rule

---

[3] See (Schroeder-Heister, 2010) for these results.

is available in the database), the argument for B continues as usual with an opponent attack on B (which must eventually be defended by the proponent), depending on the respective form of B.

We first recall the standard dialogues and winning strategies for intuitionistic logic.[4] We then introduce dialogues for implications as rules and compare them with the standard dialogues. We also discuss the inference schema of cut and the special role it takes in the implications-as-rules framework. We restrict ourselves to propositional logic throughout.

3.1 Dialogues and strategies

Our *language* consists of propositional *formulas* A, B, C, ... that are constructed from *atomic formulas* a, b, c, ... with the *logical constants* $\neg$ (negation), $\wedge$ (conjunction), $\vee$ (disjunction) and $\to$ (implication). Furthermore, $\vee$, $\wedge_1$ and $\wedge_2$ are used as *special symbols*. In addition, the letters P ('proponent') and O ('opponent') are used. An *expression* e is either a formula or a special symbol. For each expression e there is a *P-signed expression* $P\,e$ and an *O-signed expression* $O\,e$. A signed expression is called *assertion* if the expression is a formula; it is called *symbolic attack* if the expression is a special symbol. X and Y, where $X \neq Y$, are used as variables for P and O.

For each logical constant the following *argumentation forms* determine how a complex formula (having the respective logical constant in outermost position) that is asserted by X can be attacked by Y and how this attack can be defended (if possible) by X:

AF($\neg$): assertion: $X\,\neg A$
attack: $Y\,A$
defense: *no defense*

AF($\wedge$): assertion: $X\,A_1 \wedge A_2$
attack: $Y\,\wedge_i$ (Y chooses $i = 1$ or $i = 2$)
defense: $X\,A_i$

AF($\vee$): assertion: $X\,A_1 \vee A_2$
attack: $Y\,\vee$
defense: $X\,A_i$ (X chooses $i = 1$ or $i = 2$)

---

[4]We follow the presentation of Felscher (1985, 2002), with slight deviations.

Implications as Rules in Dialogical Semantics

AF($\to$): assertion: $X\,A \to B$
          attack: $Y\,A$
          defense: $X\,B$

Let $\delta(n)$, for $n \geq 0$, be a signed expression and $\eta(n)$ a pair $[m, Z]$, for $0 \leq m < n$, where Z is either A (for 'attack') or D (for 'defense'), and where $\eta(0)$ is empty. The numbers in the domain of $\delta(n)$ are called *positions*, and m in $\eta(n) = [m, Z]$ refers to a position $m < n$. Pairs $\langle \delta(n), \eta(n) \rangle$ are called *moves*. When talking about a move $\langle \delta(n), \eta(n) \rangle$, we write $\langle \delta(n) = X\,e, \eta(n) = [m, Z] \rangle$ to express that $\delta(n)$ has the form $X\,e$ and $\eta(n)$ has the form $[m, Z]$. A move $\langle \delta(n), \eta(n) = [m, A] \rangle$ is called *attack move*, and a move $\langle \delta(n), \eta(n) = [m, D] \rangle$ is called *defense move*. An attack $\langle \delta(n), \eta(n) = [m, A] \rangle$ at position n on an assertion at position m is called *open at position k* for $n < k$ if there is no position n' such that $n < n' \leq k$ and $\langle \delta(n'), \eta(n') = [n, D] \rangle$, that is, if there is no defense at or before position k to an attack at position n.

We now define a *D-dialogue* as a (possibly infinite) sequence of moves $\langle \delta(n), \eta(n) \rangle$ ($n = 0, 1, 2, \dots$) satisfying the following conditions:

(D00) $\delta(n)$ is a P-signed expression if n is even and an O-signed expression if n is odd. The expression in $\delta(0)$ is a complex formula.

(D01) If $\eta(n) = [m, A]$, then the expression in $\delta(m)$ is a complex formula and $\delta(n)$ is an attack on this formula as determined by the relevant argumentation form.

(D02) If $\eta(p) = [n, D]$, then $\eta(n) = [m, A]$ for $m < n < p$ and $\delta(p)$ is the defense of the attack $\delta(n)$ as determined by the relevant argumentation form.

(D10) If, for an atomic formula a, $\delta(n) = P\,a$, then there is an m such that $m < n$ and $\delta(m) = O\,a$. That is, P may assert an atomic formula only if it has been asserted by O before.

(D11) If $\eta(p) = [n, D]$, $n < n' < p$, $n' - n$ is even and $\eta(n') = [m, A]$, then there is a p' such that $n' < p' < p$ and $\eta(p') = [n', D]$. That is, if at a position $p - 1$ there are more than one open attacks, then only the last of them may be defended at position p.

(D12) For every m there is at most one n such that $\eta(n) = [m, D]$. That is, an attack may be defended at most once.

(D13) If m is even, then there is at most one n such that $\eta(n) = [m, A]$. That is, a P-signed formula may be attacked at most once.

Proponent P and opponent O are not interchangeable due to the asymmetries between P and O introduced in (D10) and (D13): For atomic formulas a the proponent move $\langle \delta(n) = P\,a, \eta(n) = [m, Z] \rangle$ is possible only after an opponent move $\langle \delta(m) = O\,a, \eta(m) = [k, Z] \rangle$ for $k < m < n$, and O can attack a P-signed formula only once, whereas P can attack O-signed formulas repeatedly. The argumentation forms are symmetric with respect to P and O, however, in the sense that they are independent of whether the assertion is made by P or O.

We say that P *wins a D-dialogue for a formula A* if the D-dialogue is finite, begins with the move $P\,A$ and ends with a move of P such that O cannot make another move.

A *D-dialogue tree* is a tree whose branches contain as paths all possible D-dialogues for a given formula (where a *path* in a branch of a tree with root node n_0 is a sequence $n_0, n_1, \ldots, n_k$ of nodes for $k \geq 0$ where n_i and n_{i+1} are adjacent for $0 \leq i < k$).

We define a *(winning) D-strategy* for a formula A as a subtree S of the D-dialogue tree for A such that S does not branch at even positions, S has as many nodes at odd positions as there are possible moves for O, and all branches of S are D-dialogues for A won by P.

To give an example, the following is a D-strategy for the formula $a \to ((a \to (b \wedge c)) \to b)$:

0.	$P\,a \to ((a \to (b \wedge c)) \to b)$	
1.	$O\,a$	$[0, A]$
2.	$P\,(a \to (b \wedge c)) \to b$	$[1, D]$
3.	$O\,a \to (b \wedge c)$	$[2, A]$
4.	$P\,a$	$[3, A]$
5.	$O\,b \wedge c$	$[4, D]$
6.	$P\,\wedge_1$	$[5, A]$
7.	$O\,b$	$[6, D]$
8.	$P\,b$	$[3, D]$

(In this example the D-strategy consists in only one D-dialogue, which is not necessarily the case in general.)

3.2 Dialogues for implications as rules

Now we introduce dialogues for the implications-as-rules approach. Its guiding idea is the following: When making an assertion A, the proponent P must be prepared to either defend A in the 'standard' way against an attack of the opponent O, or else make the assertion C for some C, for which O has already claimed $C \to A$, that is, for which the implication-as-rule $C \to A$ is sufficient to generate A. This is modelled by saying that every assertion of P is symbolically questioned by O, following which P chooses which of the two ways described P is prepared to take. Contrary to P, O is not given a choice. O's non-implicational assertions are attacked and defended as usual. O's implicational assertions are considered as providing rules which P can *use*, but not question; so there are no attacks and defenses defined for them.

We first define *argumentation forms* for each logical constant that determine how a complex formula that has been asserted by the opponent O can be attacked and how this attack can be defended:

AF($\neg\vdash$): assertion: $O \neg A$
  attack: $P\, A$
  defense: *no defense*

AF($\wedge\vdash$): assertion: $O\, A_1 \wedge A_2$
  attack: $P \wedge_i$ (P chooses $i = 1$ or $i = 2$)
  defense: $O\, A_i$

AF($\vee\vdash$): assertion: $O\, A_1 \vee A_2$
  attack: $P \vee$
  defense: $O\, A_i$ (O chooses $i = 1$ or $i = 2$)

AF($\to\vdash$)°: assertion: $O\, A \to B$
  attack: *no attack*
  defense: *no defense*

Except for AF($\to\vdash$)°, these argumentation forms coincide with the standard ones in case of assertions made by the opponent O.

We extend our language by the two special symbols ? and $|\cdot|$. For assertions made by the proponent P there is a pair of argumentation forms for each logical constant (depicted below as trees having two branches which are separated by $|$). An assertion A made by P can be questioned by the opponent with the move O ?. The proponent P can then answer this question either by allowing an attack on the assertion (this is indicated by the special symbol $|\cdot|$; see the argumentation forms on the left side of $|$ below), or by asserting any C for which O has asserted $C{\rightarrow}A$ at an earlier position. We call this the *rule condition R*:

(R) P may answer a question O? on a formula A by choosing C provided O has asserted the formula $C \rightarrow A$ before.

Then the argumentation forms for assertions made by P are as follows:

AF($\vdash \neg$):   assertion: $P \neg A$
          question: O?
          choice: $P|\neg A|$ $\quad\quad$ $P C$ $\quad$ (R)
          attack: $O A$
          defense: *no defense*

AF($\vdash \wedge$):   assertion: $P A_1 \wedge A_2$
          question: O?
          choice: $P|A_1 \wedge A_2|$ $\quad\quad$ $P C$ $\quad$ (R)
          attack: $O \wedge_i$ $(i = 1$ or $2)$
          defense: $P A_i$

AF($\vdash \vee$):   assertion: $P A_1 \vee A_2$
          question: O?
          choice: $P|A_1 \vee A_2|$ $\quad\quad$ $P C$ $\quad$ (R)
          attack: $O \vee$
          defense: $P A_i$ $(i = 1$ or $2)$

AF($\vdash \rightarrow$):   assertion: $P A \rightarrow B$
          question: O?
          choice: $P|A \rightarrow B|$ $\quad\quad$ $P C$ $\quad$ (R)
          attack: $O A$
          defense: $P B$

In the case of an attack $O \wedge_i$ according to the argumentation form AF($\vdash \wedge$) the opponent O chooses $i = 1$ or $i = 2$, and in the case of a defense $P A_i$ to an attack $O \vee$ according to the argumentation form AF($\vdash \vee$) the proponent P chooses $i = 1$ or $i = 2$. The argumentation forms on the left (i.e., the respective left branches) correspond to the argumentation forms of *D*-dialogues (where the device of question and choice moves is not needed). The argumentation forms on the right (i.e., the respective right branches) reflect the implications-as-rules view.

For assertions of atomic formulas a made by the proponent P an

Implications as Rules in Dialogical Semantics 219

argumentation form is given by the rule condition (R) itself:

AF(R): assertion: $P\,a$
question: $O\,?$
choice: $P\,C$ only if O has asserted $C \to a$ before

In addition, we define an argumentation form AF(Cut) such that any expression e (i.e., question, symbolic attack or formula) stated by O can be followed by a move $P\,A$, which can then be followed by the move $O\,A$, for any *cut formula* A:

AF(Cut): statement: $O\,e$
cut: $P\,A$
cut: $O\,A$

This argumentation form differs from the others in that the move $O\,e$ need not be an assertion (i.e. the statement of a formula) but can be the statement of any expression e (i.e., question, symbolic attack or formula). Another difference is that the cut formula is completely independent of the expression e. Calling the P-move an attack and the subsequent O-move a defense as in the other argumentation forms would thus be inadequate. We therefore simply speak of *cut moves* in both cases. The idea behind cut is that at any (even) position, instead of proceeding in the original way, P can introduce an arbitrary formula A as a lemma. P must then later be prepared both to defend this lemma A as an assertion and to defend his original claim *given* this lemma, that is, given the opponent's claim of A.

Formally, we extend the definition of *moves*: For $\delta(n)$ being a signed expression and $\eta(n)$ being a pair $[m, Z]$ for $0 \leq m < n$, Z is now either A (for 'attack'), D (for 'defense'), Q (for 'question'), C (for 'choice') or *Cut*. As before, pairs $\langle \delta(n), \eta(n) \rangle$ are called *moves*, where $\eta(m)$ is empty for $m = 0$ and in case of *Cut*. We have thus the following types of moves:

attack move $\langle \delta(n) = X\,e, \eta(n) = [m, A] \rangle$,
defense move $\langle \delta(n) = X\,A, \eta(n) = [m, D] \rangle$,
question move $\langle \delta(n) = O\,?, \eta(n) = [m, Q] \rangle$,
choice move $\langle \delta(n) = P\,|A|, \eta(n) = [m, C] \rangle$,
  $\langle \delta(n) = P\,A, \eta(n) = [m, C] \rangle$,
cut move $\langle \delta(n) = X\,A, \eta(n) = [Cut] \rangle$.

(A question move can only be made by O and a choice move can only be made by P. The other types of moves are available for both the proponent P and the opponent O.)

A *D°-dialogue*, which is a dialogue based on the implications-as-rules view plus cut, is now defined as a sequence of moves $\langle \delta(n), \eta(n) \rangle$ ($n = 0, 1, 2, \ldots$) satisfying the following conditions:

($D00°$) $\delta(n)$ is a P-signed expression if n is even and an O-signed expression if n is odd. The expression in $\delta(0)$ is a (complex or atomic) formula.

($D01°$) If $\eta(n) = [m, A]$, then for $m < n$ the expression in $\delta(m)$ is a complex formula for even n, or, for odd n, the expression is of the form $|B|$ for a complex formula B. In both cases $\delta(n)$ is an attack as determined by the relevant argumentation form.

($D02$) is the same as above.

($D03°$) If $\eta(n) = [m, Q]$ (for odd n), then for $m < n$ the expression in $\delta(m)$ is a (complex or atomic) formula, $\eta(m) = [l, Z]$ for $l < m$, $Z = A, D, C$ or Cut (where l is empty if $Z = Cut$), and the expression in $\delta(n)$ is the question mark '?'.

($D04°$) If $\eta(n) = [m, C]$ (for even n), then $\eta(m) = [l, Q]$ for $l < m < n$ and $\delta(n)$ is the choice answering the question $\delta(m)$ as determined by the relevant argumentation form.

($D05°$) If $\eta(n) = [Cut]$ for even n, then $\eta(m) = [l, Z]$ (where l is empty if $Z = Cut$) for $l < m < n$ and $\delta(n)$ is a formula (i.e. the cut formula). If $\eta(n) = [Cut]$ for odd n, then $\eta(n-1) = [Cut]$ and $\delta(n) = O\,A$ for $\delta(n-1) = P\,A$.

($D11$) and ($D12$) are the same as above.

($D13°$) If m is even, then there is at most one n such that $\eta(n) = [m, Z]$ for $Z = Q$ or $Z = A$. That is, a P-signed formula, resp. a P-signed expression of the form $|B|$, may be questioned, resp. attacked, at most once.

($D14°$) O can question a formula C if and only if (i) C has not yet been asserted by O, or (ii) C has already been attacked by P.

The notions 'dialogue won by P', 'dialogue tree' and 'strategy' as defined for D-dialogues are directly carried over to the corresponding notions for $D°$-dialogues.

The conditions defining $D°$-dialogues are similar to those defining D-dialogues. Two important differences are the absence of condition ($D10$) and the additional condition ($D14°$) in the former. The absence of ($D10$) is compensated for by allowing O to question assertions of

Implications as Rules in Dialogical Semantics 221

atomic formulas made by P, and by the presence of $(D14°)$. Condition $(D00°)$ allows $D°$-dialogues to start with the assertion of an atomic formula, contrary to the restriction to complex formulas in D-dialogues. Conditions $(D03°)$ and $(D04°)$ have been added for the question and choice moves, respectively, and condition $(D05°)$ has been added for the cut moves. Note that by $(D05°)$ the opponent O can make a cut move only immediately after a cut move made by P.

For example, a $D°$-strategy for the formula $a \to ((a \to (b \wedge c)) \to b)$ is the following (for comparison, see the above D-strategy for this formula):

0.	$P\, a \to ((a \to (b \wedge c)) \to b)$			
1.	$O\, ?$			$[0, Q]$
2.	$P\, \vert a \to ((a \to (b \wedge c)) \to b)\vert$			$[1, C]$
3.	$O\, a$			$[2, A]$
4.	$P\, (a \to (b \wedge c)) \to b$			$[3, D]$
5.	$O\, ?$			$[4, Q]$
6.	$P\, \vert (a \to (b \wedge c)) \to b\vert$			$[5, C]$
7.	$O\, a \to (b \wedge c)$			$[6, A]$
8.	$P\, b \wedge c$			$[Cut]$
9.	$O\, ?$	$[8, Q]$	$O\, b \wedge c$	$[Cut]$
10.	$P\, a$	$[9, C]$	$P\, \wedge_1$	$[9, A]$
11.			$O\, b$	$[10, D]$
12.			$P\, b$	$[7, D]$

The moves at positions 0–4 and at positions 4–7 + 12 (in the right dialogue) are made according to the argumentation form $AF(\vdash \to)$. In the choice moves at positions 2 resp. 6 the proponent P can only choose $\vert a \to ((a \to (b \wedge c)) \to b)\vert$ resp. $\vert (a \to (b \wedge c)) \to b\vert$, since O has not asserted any implications before that could be used as rules by choosing their antecedents. This is different in the choice move at position 10 (in the left dialogue): The opponent O has claimed the implication $a \to (b \wedge c)$ before at position 7, whose succedent is exactly the formula asserted by P at position 8, which is questioned by O at position 9. The proponent P can now use this implication as a rule by answering the question on $b \wedge c$ with the assertion of its antecedent a in the choice move at position 10. This assertion cannot be questioned further due to condition $(D14°)$; likewise for the assertion of b at position 12. Hence both dialogues are won by P, and we have a $D°$-strategy.

It can be shown that there is no $D°$-strategy without cut moves for the formula $a \to ((a \to (b \wedge c)) \to b)$. The above D°-strategy corresponds to the LI°-derivation (1). Furthermore, it can be shown that the weak cut elimination property also holds for D°-strategies. That is, every D°-strategy containing cut moves can be transformed into a D°-strategy of the form[5]

$$
\begin{array}{ll}
& \vdots \\
m. & O\ A \to B\ [m-1, Z] \\
& \vdots \\
n. & P\ B\ [Cut] \\
n+1. & O?\ [n, Q] \quad\bigg|\quad O\ B\ [Cut] \\
n+2. & P\ A\ [n+1, C] \quad\ \ s_2 \\
n+3. & O?\ [n+2, Q] \\
& s_1
\end{array}
$$

where the O-move at position m is either an attack or a defense (i.e., either $Z = A$ or $Z = D$), and the move $\langle \delta(n+1) = O\ B, \eta(n+1) = [Cut]\rangle$ is the uppermost cut move made by O (i.e., there is no cut move at positions $k < n-1$). The O-move at position $n+3$ might not be possible due to $(D14^\circ)$. In this case the left dialogue ends with the P-move at position $n+2$. Moreover, the substrategy s_2 is either of the same form as the above D°-strategy, or it depends on a sequence of moves made according to AF($\neg\vdash$), AF($\wedge\vdash$), AF($\vee\vdash$) or AF($\to\vdash$)$^\circ$. This corresponds to the properties of LI°-derivations (cf. the LI°-derivation (2) above).

It can be shown that the sequent calculus LI° is sound and complete with respect to the dialogical semantics given by D°-dialogues.

4 Discussion

We have presented a Lorenzen-style dialogue framework for the interpretation of implications as rules which is equivalent to the sequent calculus LI° incorporating this interpretation. The dialogical framework is not as straightforward as LI°, which can be read as the proof-theoretic semantics for implications as rules. Does this speak against the dialogical approach, or perhaps against the idea of implications as rules?

What makes the dialogical presentation difficult to grasp at first

---

[5] Where the moves at positions m, $n+1$, $n+2$ and $n+3$ can even be assumed to refer to the immediately preceding moves, respectively.

Implications as Rules in Dialogical Semantics 223

sight is that the usual symmetry between proponent and opponent is lost. Although P and O play different roles in any Lorenzen-style dialogue game, with respect to the attack and defense principles we normally have a perfect symmetry. Just attacks and defenses are defined, not different ways of attacking and defending for P or O. This idea is so deeply rooted in the dialogical paradigm that giving it up may appear as giving up the dialogical setting itself as a foundational approach. The counterargument from the implications-as-rules view would be that implication is different from the other connectives, and that this difference requires an asymmetric treatment. If one wants to formally keep symmetry for implication as a logical connective, one could distinguish between implications $A \to B$ and rules $A \Rightarrow B$, and reduce implications to rules by separate inferences. An attack on an implication $A \to B$ would be defended by claiming a rule $A \Rightarrow B$. Asymmetry would only come in for the rule $A \Rightarrow B$ considered as a 'structural entity', not yet for the implication $A \to B$. This way of proceeding involves, of course, some duplication of notation.

The asymmetry in the treatment of implication brings another asymmetry with it: The proponent can now defend a proposition A by means of the *rule condition* independent of the logical form of A, as an alternative to the 'standard' defense of A which depends on its logical form. This possibility is open only to the proponent and does not fit into the dialogical schema which decomposes formulas according to their logical form.

However, principles of decomposition and symmetry should not be taken as sacrosanct, in particular as rules are *not* logical constants but belong to the general structural framework on top of which logical constants are defined. Given that P has the dialogical role of claiming something to hold, and O the role of providing the assumptions under which something is supposed to hold, the rule $A \Rightarrow B$ means for P that B must be defended on the background A, whereas O only grants with $A \Rightarrow B$ the right to *use* it as a rule, without any propositional claim. This is exactly what is expressed in the dialogue rules for implications-as-rules presented in this paper.

A crucial aspect here is the significance which is given to modus ponens. For the implications-as-rules view, modus ponens is essential for the meaning of implication as it expresses the idea of *application*, which is the characteristic feature of a rule. In a natural-deduction setting with rules made explicit, the application of a rule $A_1, \ldots, A_n \Rightarrow B$

is framed as a generalized modus ponens, which, when applied to premisses $A_1,\ldots,A_n$, yields the conclusion B (Schroeder-Heister, 2012). The system LI° can be viewed as a calculus representing the idea of modus ponens at the sequent-calculus level. The standard interpretation of implication in the dialogical setting corresponds instead to the symmetric sequent calculus LI which is based on the 'implications-as-links' view. According to this view, an implication $A \to B$ which is introduced on the left side of the sequent sign by means of Gentzen's implication-left schema

$$(\to \vdash) \frac{\Gamma \vdash A \quad B, \Delta \vdash C}{\Gamma, A \to B, \Delta \vdash C}$$

links an occurrence of A on the right side of the left premiss with an occurrence of B on the left side of the right premiss of this rule.

The standard dialogical approach favours sequent-style reasoning in the sense of $(\to \vdash)$. We have shown that natural-deduction style reasoning, into which the idea of implications-as-rules fits very neatly, and which can be given a sequent-style rendering via LI°, can be fully represented in the dialogical setting. This representation has the price that implications-as-rules receive an asymmetric treatment, which ultimately reflects differences between natural deduction and the symmetric sequent calculus LI rather than deficiencies of the dialogical setting or of the system LI° being modelled.

This situation is slightly complicated by the presence of cut. In order to achieve full deductive power, the presence of implications-as-rules required the use of (restricted) cut as a primitive rule. In the natural-deduction setting this is easily accommodated, as conclusions of applications of assumption rules can be premisses of elimination rules without creating a maximum formula. In the dialogical setting the handling of cut is difficult and by far not as plausible as in proof systems, since one has to model the claim of the cut formula by P and O according to the pattern of attack and defense. It should be remarked, however, that in a general natural-deduction setting with rules of arbitrary levels and general principles of definitional reflection, it might be reasonable to use a weaker notion of rule without the presupposition of cut, so that this problem disappears at the general level[6].

---

[6]This is investigated in forthcoming work by Lars Hallnäs and the second author.

Overall this paper demonstrates again that the dialogical framework is versatile enough to deal with approaches originally developed in the realm of proof-theoretic semantics. In the end, more general arguments are needed if one wants to give preference either to proofs or to dialogues as the appropriate foundational approach.

References

de Campos Sanz, W., & Piecha, T. (2011). A criticism of the BHK interpretation. *Bulletin of Symbolic Logic*, *17*, 292. (Abstract for the ASL Logic Colloquium 2010, Paris, July 25–31.)

Felscher, W. (1985). Dialogues, strategies, and intuitionistic provability. *Annals of Pure and Applied Logic*, *28*, 217–254.

Felscher, W. (2002). Dialogues as a foundation for intuitionistic logic. In D. M. Gabbay & F. Guenthner (Eds.), *Handbook of philosophical logic* (2nd ed., Vol. 2, pp. 115–145). Dordrecht: Kluwer.

Heyting, A. (1971). *Intuitionism. An introduction* (3rd ed.). Amsterdam: North-Holland.

Lorenzen, P. (1960). Logik und Agon. In *Atti del XII congresso internazionale di filosofia (Venezia, 12–18 Settembre 1958)*. (Vol. quarto, pp. 187–194). Firenze: Sansoni Editore.

Schroeder-Heister, P. (2010). Implications-as-rules vs. implications-as-links: An alternative implication-left schema for the sequent calculus. *Journal of Philosophical Logic*, *40*, 95–101.

Schroeder-Heister, P. (2012). Generalized elimination inferences, higher-level rules, and the implications-as-rules interpretation of the sequent calculus. In E. H. Haeusler, L. C. Pereira, & V. de Paiva (Eds.), *Advances in natural deduction*.

Sørensen, M. H., & Urzyczyn, P. (2006). *Lectures on the Curry-Howard isomorphism*. New York: Elsevier.

Thomas Piecha and Peter Schroeder-Heister
Wilhelm-Schickard-Institut für Informatik, Universität Tübingen
Sand 13, 72076 Tübingen, Germany
e-mail: piecha@informatik.uni-tuebingen.de,
   psh@uni-tuebingen.de
URL: http://ls.inf.uni-tuebingen.de

Archetypal Rules and Intermediate Logics

Tomasz Połacik

Abstract

The notions of archetypal rule and universally representative connective were introduced by L. Humberstone. We study these notions in the context of intermediate logics and present three characterizations of classical propositional logics in terms of the notions in question.

Keywords: classical propositional logic, intermediate propositional logics, rules of inference

*There is still much good music
to be written in C major.*
A. Schoenberg

1 Introduction

Probably one can hardly believe that there are still interesting problems concerning classical propositional logic. But at least, one can always find a valuable inspiration in investigation of this simple case. A good example of such an inspiration can be found in (Humberstone, 2004), where the Author introduces the notion of archetypal rule and considers it in the context of classical and intuitionistic propositional logic.

In this paper we consider intermediate logics, i.e., propositional logics that are contained in classical logic, CPC, and contain intuitionistic logic, IPC. It is well known that the intermediate logics with a natural ordering form a lattice with intuitionistic logic at the bottom and classical logic at the top. A continuum of logics between intuitionistic and classical logic create an interesting field of investigations. For more information and further reference, consult e.g. (Chagrov & Zakharyaschev, 1997) and (Pogorzelski & Wojtylak, 2008).

Typically, given a property that may, or may not, be enjoyed by the logics in question, it is natural to investigate the distribution of this property in the whole lattice. Let us give a few examples.

Recall that we say that an intermediate logic L enjoys disjunction property if, whenever L proves the disjunction of two formulas, then one of its disjuncts is also provable in L. The disjunction property may be regarded as a key feature of constructive logics. One can clearly see that CPC rejects disjunction property (it is sufficient to think of the principle of excluded middle) while it is validated by IPC. So, the natural question arises: where is the borderline between logics with and logics without the disjunction property, or between logics that are constructive and non-constructive ones? J. Łukasiewicz conjectured that IPC is the only intermediate logic that possesses disjunction property. However, this conjecture turned out to be false. It was proven by G. Kreisel and H. Putnam by pointing to the logic with disjunction property stronger that IPC. Now we know that disjunction property is non-trivially distributed over the lattice of intermediate logics. For example, there is no greatest intermediate logic having the disjunction property, and there is a continuum of maximal logics which enjoy disjunction property.

Some interesting properties of intermediate logics can be expressed in terms of rules of inference. Before we give an example, recall that an inference rule r is called admissible for a logic L if, whenever all the the premises of r are derivable in L, so is the conclusion of r. A relatively recent example of exploration of the lattice of intermediate logics concerns the so-called basis for all admissible rules and is due to R. Iemhoff. Here, a natural countable basis $\mathcal{V}$ for the admissible rules of IPC is found and it is shown that there is no proper intermediate logic with the disjunction property for which all rules in $\mathcal{V}$ are admissible. It follows that IPC is the only intermediate logic with the disjunction property which has this property.

Recall also that we say that a rule r is derivable in an intermediate logic L if the conclusion of r can be derived in L from the premises of r. Obviously, for every intermediate logic L, every derivable rule is admissible, but not necessarily the opposite. If, however these two sets of rules coincide for some logic L, then L is Post-complete. The notion of structural completeness refines that of Post completeness.

Namely, we say that a logic L is structurally complete if, restricting to structural rules of L only, all its admissible rules are derivable. It is known that the property of structural completeness is enjoyed by CPC, and IPC is not structurally complete. Thus, the problem of distribution of the property of being structurally complete arises in a natural way. It turns out that the property of being structurally complete does not characterize CPC. Among known examples of logics that are structurally complete, Medvedev's Logic, ML, seems to be particularly attractive. This is the only *known* intermediate logic which is structurally complete and possesses disjunction property. The longstanding open problem is whether ML is actually characterized by these two properties.

In this paper we focus on the notion of archetypal rule introduced in (Humberstone, 2004). We consider some properties which can be expressed by means of archetypality and related notions and investigate the problem of distribution of these properties in the lattice of intermediate logics.

2 Archetypal rules

Let us begin with an introduction of basic notions. We consider the usual language of propositional logic containing the propositional connectives $\neg, \wedge, \vee, \rightarrow$ and $\leftrightarrow$ and the constants $\bot$ and $\top$. A rule, whose premisses are $A_1, \ldots, A_n$ and B is the conclusion, will be denoted by $A_1, \ldots, A_n/B$. Since any inference rule with a finite number of premisses can be represented in our language as a one-premiss rule, we confine ourselves only to such rules, denoted here by expressions of the form A/B. We say that a rule A/B is *n-ary* (*binary*) if its premiss and conclusion involve at most n (at most two) propositional variables.

Let $\vdash_L$ be a consequence relation for a given intermediate logic L. As usual, we say that a rule A/B is *derivable* according to $\vdash_L$ if $\sigma(A) \vdash_L \sigma(B)$, for any substitution σ for the propositional variables of the formulas A and B.

Let us paraphrase some ideas of (Humberstone, 2004) within the following definition.

Definition 1 A rule A/B is *non degenerate in* the logic L iff it is

- non-overflowing, i.e., $A \nvdash_L \bot$,

- non-trivial, i.e., $\not\vdash_\mathsf{L} B$,
- non-reversible, i.e., $B \not\vdash_\mathsf{L} A$.

A rule which is non-degenerate and derivable will be called *normal*.

Note In the sequel, if it is not explicitly stated otherwise, by a *rule* we will always mean a *derivable rule*.

Let us illustrate our basic concepts with the following two examples borrowed from Humberstone (2004).

Consider any intermediate logic L and the rule $R = p \wedge q/q$. It is easy to see that R is a normal rule in L. Now, let us consider any other derivable rule, not necessarily non-degenerate one, $S = C/D$ and think of the substitution σ such that $\sigma(p) = C$ and $\sigma(q) = D$. Then, one can easily see that

$$\vdash_\mathsf{L} \sigma(p \wedge q) \leftrightarrow C \qquad \vdash_\mathsf{L} \sigma(q) \leftrightarrow D.$$

Thus, modulo provability in the logic L, every derivable rule can be obtained as a substitutional instance of the rule R.

Furthermore, let us consider another rule, $R' = \neg p/\neg p \vee q$. It should be clear that the rule R' is normal in every intermediate logic. Let us think of a rule $S' = E/F$ and a substitution σ' such that $\sigma'(p) = \neg E$ and $\sigma'(q) = F$. Now, we can easily check that

$$\vdash_\mathsf{CPC} \sigma'(\neg p) \leftrightarrow E \quad \text{and} \quad \vdash_\mathsf{CPC} \sigma'(\neg p \vee q) \leftrightarrow F.$$

Thus, modulo provability *in classical propositional logic* CPC, every derivable rule can be obtained as a substitutional instance of the rule R'. Note that this is not longer true in case of IPC.

One can think that, in some sense, the rule R defined above is the most general one for every intermediate logic, since any other derivable rule is just an substitutional instance of the rule R. On the other hand, the rule R' might be regarded by the most general rule in classical logic, but not in intuitionistic logic. This observation motivates introducing the following definition.

Definition 2 (Humberstone, 2004) A rule A/B is called *archetypal for a logic* L iff for every rule A'/B' there is a substitution σ, such that

$$\vdash_\mathsf{L} \sigma(A) \leftrightarrow A' \quad \text{and} \quad \vdash_\mathsf{L} \sigma(B) \leftrightarrow B'.$$

Archetypal Rules and Intermediate Logics 231

In (Humberstone, 2004) we can find many examples of rules which are archetypal or not archetypal. Let us just mention a few. Firstly, the rules $p \wedge q/p$, $p/p \vee q$ and $p/(p \to q) \to q$ are archetypal for every intermediate logic. On the other hand, the rules $p/p \vee \neg p$, $p/\neg p \to q$ or even $p \leftrightarrow q/p \to q$ are not archetypal for intuitionistic logic. Notice that the latter example is a substitutional instance of the conjunction-elimination rule which, as we have already seen, is archetypal for every intermediate logic.

The problem of archetypality turns up to be rather complex in case of intuitionistic logic. Probably there are many interesting problems concerning archetypality in other intermediate logics as well. However, one can expect a clear picture of the properties of archetypal rules in classical logic. Indeed, L. Humberstone proved the following result.

Theorem 1 (Humberstone, 2004) *In classical propositional logic, every **binary** normal rule is archetypal.*

One could expect that this result can be generalized to the whole class of normal rules. However, due to specific methods used in the proof, it cannot be proven by a simple generalization of the original proof.

The expected generalization of Theorem 1 can be found in (Połacik, 2005). The main idea of the proof presented there is to reduce, via a suitable substitution, an arbitrary normal rule to a binary one and then apply Humberstone's Theorem. Note that if a rule A/B is derivable then so is the rule $\tau A/\tau B$, for every substitution τ. Unfortunately, the latter rule need not be non-degenerate, even if the former is. Thus the important point in proving the theorem is the fact that, with some care, one can also preserve non-degeneracy of the rule.

Here we present the generalized theorem of (Połacik, 2005) with another shorter algebraic proof.

Theorem 2 *In classical propositional logic, for every normal rule A/B there is a substitution τ such that $\tau A/\tau B$ is a normal **binary** rule.*

Proof. Let $\mathcal{A}_n$ be the Lindenbaum algebra of classical propositional logic in the language with propositional variables $\{p_1, \ldots, p_n\}$ where $n \geq 2$. Notice that $\mathcal{A}_n$ is a finite boolean algebra with exactly 2^n atoms. Let $\leq$ stand for the usual ordering of the algebra $\mathcal{A}_n$ and let $[A]$ stand for for the equivalence class of a formula A.

Assume that A/B is a normal rule where $A = A(\vec{p})$ and $B = B(\vec{p})$, where $\vec{p} = p_1, \ldots, p_n$ and $n \geq 2$. Then, in the Lindenbaum algebra $\mathcal{A}_n$ we have

$$[A] \leq [B], \qquad (1)$$

since A/B is derivable, and

$$0 < [A], \qquad [B] < 1, \qquad [B] \not\leq [A], \qquad (2)$$

since A/B is non-degenerate.

So, according to (2), we can find three atoms a, b, c of $\mathcal{A}_n$ such that

$$a \leq [A] \qquad b \leq [\neg B], \qquad c \leq [B], \qquad c \leq [\neg A]. \qquad (3)$$

Recall that there are 2^n atoms in $\mathcal{A}_n$ and $n \geq 2$. So, we can find another atom d of $\mathcal{A}_n$, distinct from a, b and c. On the other hand let w, x, y, z be the four atoms of the Lindenbaum algebra $\mathcal{A}_2$. Let us consider the following mapping from the set of atoms of $\mathcal{A}_n$ to the set $\{w, x, y, z\}$ of all atoms of $\mathcal{A}_2$:

$$f(u) = \begin{cases} w & \text{if } u = a, \\ x & \text{if } u = b, \\ y & \text{if } u = c, \\ z & \text{if } u = d, \\ 0 & \text{for other atoms} \end{cases}$$

Notice that all the atoms of the algebra $\mathcal{A}_2$ are in the range of f. So, since every element of a finite boolean algebra is uniquely determined as the join of all atoms that are "below" it, the mapping f can be uniquely extended to a homomorphism from $\mathcal{A}_n$ onto $\mathcal{A}_2$.

This homomorphism, in turn, uniquely determines the substitution τ for propositional variables. Obviously, the inequalities (1) and (3) are preserved by the homomorphism in question, i.e.,

$$[\tau(A)] \leq [\tau(B)], \qquad (4)$$

and

$$\begin{array}{ll} 0 < w \leq [\tau(A)] & 0 < x \leq [\tau(\neg B)] \\ 0 < y \leq [\tau(B)] & 0 < y \leq [\tau(\neg A)]. \end{array} \qquad (5)$$

So, it follows from (4) and (5) that $\tau(A)/\tau(B)$ is a binary normal rule in CPC. $\square$

Archetypal Rules and Intermediate Logics

Theorem 2 allows us to reduce the general problem of archetypality in CPC to the case of binary rules and Humberstone's Theorem. Indeed, if A/B is a normal rule, then there is a substitution τ such that $\tau(A)/\tau(B)$ is a binary normal rule. Now we can apply Theorem 1 and conclude that rule $\tau(A)/\tau(B)$ is archetypal. So, for another derivable rule C/D there is a substitution σ such that $\vdash_{\mathsf{CPC}} C \leftrightarrow \sigma\tau(A)$ and $\vdash_{\mathsf{CPC}} D \leftrightarrow \sigma\tau(B)$ and this means that the rule A/B itself is archetypal. Thus, we have proved the following fact.

Theorem 3 *In classical propositional logic, every normal rule is archetypal.*

Let us illustrate our result by some examples.

Example 1 Let us consider the following 4-ary rule R:
$$\frac{((p \to q) \to t)}{((t \to p) \to (s \to p))}.$$
One can easily check that R is a normal rule in CPC. Now, if we define the substitution σ as
$$\sigma(p) = \sigma(q) = \sigma(t) := x \text{ and } \sigma(s) := y,$$
then it can be seen that
$$\vdash_{\mathsf{CPC}} \sigma((p \to q) \to t) \leftrightarrow x$$
$$\vdash_{\mathsf{CPC}} \sigma((t \to p) \to (s \to p)) \leftrightarrow (y \to x)$$
Thus, the rule R can be reduced, via σ, to the rule
$$\frac{x}{y \to x}.$$
But the latter rule is a binary normal rule in CPC.

Example 2 Similarly as before we can show that the rule S given by
$$\frac{p \to q}{(q \to s) \to (p \to s)}$$
can be reduced, via the substitution τ such that
$$\tau p := x, \ \tau q = \bot \text{ and } \tau s := y,$$
to the rule
$$\frac{\neg x}{x \to y},$$
which is a binary normal rule in CPC.

3 Characterizations

Let say that an intermediate logic L has the property $\mathcal{H}$ if all its normal rules are archetypal. What can be said about the distribution of the property $\mathcal{H}$ in the lattice of intermediate logics? Assume that an intermediate logic L enjoys the property $\mathcal{H}$. Let us consider the rules of the form $\neg\neg p/A$ and p/B. Since one of these rules must be an substitutional instance of the other, we arrive to the conclusion that $\vdash_\mathsf{L} \tau(\neg\neg p) \leftrightarrow p$, for some substitution τ. But this may happen if and only if L is inconsistent or $\vdash_\mathsf{L} p \to \neg\neg p$. So, in particular, L = CPC. Thus, we get the following characterization of classical propositional logic, (cf. Połacik, 2005).

Theorem 4 *Classical propositional logic is the unique intermediate logic whose every normal rule is archetypal.*

In the sequel, we will present two more characterizations of classical propositional logic. Firstly, we consider the so-called universally representative propositional connectives. First, let us recall a definition due to L. Humberstone.

Definition 3 An n-ary propositional connective $P(p_1, \ldots, p_n)$ is called *universally representative* in an intermediate logic L, iff for every formula A there are formulas $A_1, \ldots, A_n$ such that

$$\vdash_\mathsf{L} A \leftrightarrow P(A_1, \ldots, A_n).$$

The fact that every propositional connective is universally representative in classical propositional logic was mentioned without the proof in (Humberstone, 2004). Below we state the theorem and sketch our proof.

Theorem 5 *Every non-constant connective P is universally representative in* CPC.

Proof. Consider an n-ary propositional connective $P(\vec{p})$, where $n > 0$ and $\vec{p} = p_0, \ldots, p_{n-1}$. We will show that for arbitrary formula $C(\vec{q})$ where $\vec{q} = q_0, \ldots, q_{m-1}$ and $m > 0$, there is a substitution $\sigma \colon \vec{p} \to \vec{q}$ such that for every $p \in \vec{p}$ we have $\sigma(p) \in \{C(\vec{q}), P(C(\vec{q}), \ldots, C(\vec{q}))\}$ and

$$\vdash_\mathsf{CPC} C(\vec{q}) \leftrightarrow P(\sigma(p_1), \ldots, \sigma(p_n)). \tag{6}$$

Let $\vec{p_0} = p_0, \ldots, p_0$. We show that

$$\vdash_{\mathsf{CPC}} p_0 \leftrightarrow P(A_0, \ldots, A_n) \text{ where } A_i = p_0 \text{ or } A_i = P(\vec{p_0}). \quad (7)$$

There are four possible cases to be considered.

Case 1 $\vdash_{\mathsf{CPC}} P(\vec{\bot}) \leftrightarrow \bot$ and $\vdash_{\mathsf{CPC}} P(\vec{\top}) \leftrightarrow \bot$.

Then we have $\vdash_{\mathsf{CPC}} P(\vec{p_0}) \leftrightarrow \bot$. Since P is non-constant there is a sequence $x_0, \ldots, x_n$ of $\top$'s and $\bot$'s that $\vdash_{\mathsf{CPC}} P(x_0, \ldots, x_n) \leftrightarrow \top$. We put

$$A_i = \begin{cases} p_0 & \text{if } x_i = \top, \\ P(\vec{p_0}) & \text{if } x_i = \bot. \end{cases}$$

Case 2 $\vdash_{\mathsf{CPC}} P(\vec{\bot}) \leftrightarrow \bot$ and $\vdash_{\mathsf{CPC}} P(\vec{\top}) \leftrightarrow \top$.

Then $\vdash_{\mathsf{CPC}} P(\vec{p_0}) \leftrightarrow p_0$ and we put $A_i = p_0$.

Case 3 $\vdash_{\mathsf{CPC}} P(\vec{\bot}) \leftrightarrow \top$ and $\vdash_{\mathsf{CPC}} P(\vec{\top}) \leftrightarrow \bot$.

Then $\vdash_{\mathsf{CPC}} P(\vec{p_0}) \leftrightarrow \neg p_0$ and we take $A_i = P(\vec{p_0})$.

Case 4 $\vdash_{\mathsf{CPC}} P(\vec{\bot}) \leftrightarrow \top$ and $\vdash_{\mathsf{CPC}} P(\vec{\top}) \leftrightarrow \top$.

Then $\vdash_{\mathsf{CPC}} P(\vec{p_0}) \leftrightarrow \top$ and $P(y_0, \ldots, y_n) = \bot$, for some sequence of $y_0, \ldots, y_n$ of $\top$'s and $\bot$'s. Then we put

$$A_i = \begin{cases} p_0 & \text{if } y_i = \bot, \\ P(\vec{p_0}) & \text{if } y_i = \top. \end{cases}$$

In each case, for $1 \leq i \leq n$ we define $\sigma(p_i) = A_i$ and verify that (7) holds. From (7), by extensionality we get, that for every formula $C = C(\vec{q})$,

$$\vdash_{\mathsf{CPC}} C \leftrightarrow P(A_0, \ldots, A_n) \text{ where } A_i = C \text{ or } A_i = P(\vec{C}).$$

So, every formula can be rewritten as a formula with P as its main connective according to (6). □

Let us say that an intermediate logic L has the property $\mathcal{C}$ iff every propositional connective is universally representable in L. Assume that L is a such a logic and think of a connective of double negation, i.e., define $P(p) = \neg\neg p$. This implies, in particular, that $\vdash_\mathsf{L} \neg\neg p \to p$, so L must be equal to CPC. We conclude that an intermediate logic L has the property $\mathcal{C}$ iff L proves the double negation law which proves the following theorem.

Theorem 6 *Classical propositional logic is the unique intermediate logic in which every non-constant propositional connective is universally representative.*

Now we introduce a pre-order on the set of all rules in a given intermediate logic L: we put $C/D \preceq_\mathsf{L} A/B$ iff for some substitution σ, we have $\vdash_\mathsf{L} C \leftrightarrow \sigma A$ and $\vdash_\mathsf{L} D \leftrightarrow \sigma B$. In the usual way, we think of the equivalence classes of the rules defined by

$$[A/B] := \{\, C/D : C/D \preceq A/B \text{ and } A/B \preceq C/D \,\}.$$

Finally, we induce a partial ordering on the set of all equivalence classes putting

$$[A/B] \leq [C/D] : \iff A/B \preceq C/D$$

(we suppress the subscript L for better readability). Now, let us consider set $\mathcal{NR}(\mathsf{L})$ of the equivalence classes of all *normal* rules of a given intermediate logic L partially ordered by $\leq$. Finally, let us observe that the rule A/B is archetypal for L iff the element $[A/B]$ is the supremum of $\mathcal{NR}(\mathsf{L})$.

As we have mentioned before, the structure of $\mathcal{NR}(\mathsf{IPC})$ seems to be rather complex. However, the structure of $\mathcal{NR}(\mathsf{CPC})$ reveals to be very simple, since it consists of just one element. Moreover, this fact combined with Theorem 4 implies the following.

Theorem 7 *Classical propositional logic is the unique intermediate logic whose poset of normal rules consists of exactly one element.*

The notion of archetypal rule and the related notions considered in our paper allow us to give three characterizations of classical propositional logic. In conclusion, we summarize all these facts within the following corollary.

Corollary 8 (Characterizations of classical logic) *Classical propositional logic is the unique intermediate logic*

- *whose every normal rule is archetypal,*
- *whose every non-constant propositional connective is universally representable,*
- *whose poset of normal rules has exactly one element.*

References

Chagrov, A., & Zakharyaschev, M. (1997). *Modal logic*. Clarendon Press.

Humberstone, L. (2004). Archetypal forms of inference. *Synthese, 141*(1), 45–76.

Pogorzelski, W., & Wojtylak, P. (2008). *Completeness theory for propositional logics*. Birkhäuser.

Połacik, T. (2005). The unique intermediate logic whose every rule is archetypal. *The Logic Journal of IGPL, 13*(3), 269–275.

Tomasz Połacik
Institute of Mathematics, University of Silesia
Bankowa 14, 40-007 Katowice, Poland
e-mail: polacik@us.edu.pl

Semantic Paradoxes and Transparent Intensional Logic

Jiří Raclavský

Abstract

The paper describes the solution to semantic paradoxes pioneered by Pavel Tichý and further developed by the present author. Its main feature is an examination (and then refutation) of the hidden premise of paradoxes that the paradox-producing expression really means what it seems to mean. Semantic concepts are explicated as relative to language, thus also language is explicated. The so-called 'explicit approach' easily treats paradoxes in which language is explicitly referred to. The residual paradoxes are solved by the 'implicit approach' which employs ideas made explicit by the former one.

Keywords: semantic notions, semantic paradoxes

Transparent Intensional Logic (*TIL*) is a rich and powerful logical system capable to treat, *inter alia*, a great amount of natural language phenomena. The aim of this rather short paper is to describe the TIL-based approach to semantic paradoxes. Pavel Tichý, the originator of TIL, developed its core ideas when he investigated and solved four versions of the Liar paradox (Tichý, 1988, section 44). The present author has elaborated his ideas into an extensive theory in a number of writings (references are suppressed).

Semantic paradoxes (*SP*s), e.g. the Liar and Grelling's heterological paradox, are paradoxes concerning semantic concepts (e.g. truth, denotation, reference). Among their premises (since paradoxes are arguments), it always occurs a *paradox-producing term* (e.g. 'This sentence is not true') which includes some *semantic term* expressing a *semantic concept*. A valuable solution to SPs should revise (a) our *uncritical* (naïve) *theory* of semantic concepts or (b) our ordinary, uncritical derivation rules, suggesting thus a *critical theory* of semantic concepts or derivation rules.

In recent decades, the classical 'hierarchical' approaches by Russell and Tarski, and even three-(many-)valued approaches by Łukasiewicz, Kripke and others, have been repudiated in favour of rather unclassical ones: Priest's paraconsistent logic (dialetheias), Gupta's and Belnap's revision theory (circular concepts and definitions), Field's paracompleteness, and contextualism. In contrast to these recent approaches, the TIL-approach is rather classical, offering an a.-style of explanation.

Here are the corner-stones of the TIL-approach:

1. critical examination, and then refutation, of the *hidden premise* of SPs that the paradox-producing expression means what it seems to mean (generalized from Tichý, 1988, p. 228);

2. since it is a truism that an expression may mean (denote, refer to) something only relative to a particular language, semantic concepts are explicated as inescapably *relative to language* (especially in Raclavský, 2009a), thus also the concept of *language* is explicated (Raclavský, 2009a).

The paper is divided into three parts. The section 1 *TIL basics* explains the notion of construction, explication of meanings (semantic scheme), and the TIL type theory. The section 2 '*Explicit approach*' provides an explication of language, explication of semantic concepts as explicitly relative to language, and a principle of solution to (many) SPs. The last section 3 '*Implicit approach*' starts with an objection, the admission of which seems to lead to the revenge problem; then, semantic concepts *implicitly* relative to language are investigated and a solution to residual SPs is explained.

1 TIL basics

1.1 Constructions

One can distinguish two notions of function: function as a mere mapping (hereafter *function*), i.e. function in 'extensional' sense, and function as a structured recipe, procedure, i.e. function in 'intensional' sense. Recall that, for instance, Russell of the no-class theory repudiated functions in the first sense while espousing functions in the letter sense (viz. his propositional functions). Tichý treats functions in both senses, the latter ones explicated as certain *constructions*. For an extensive defence of the notion of construction see (Tichý, 1988).

Constructions are structured abstract extra-language procedures (roughly: algorithms). Any object O is constructible by infinitely many *equivalent* (more precisely *v-congruent*, where v is a valuation), yet *not identical*, constructions. Two features specify each construction C: (i) which object O (if any) is (v-)constructed by C; (ii) how C (v-)constructs O (by means of which subconstructions).

For exact specification of constructions see (Tichý, 1988, ch. 5). Four basic kinds of constructions are specified there; having thus (where X is any object or construction and C_i is any construction):

i. variables x (not as letters!)

ii. trivializations ^0X ('constants')

iii. compositions $[C\, C_1 \ldots C_n]$ ('applications')[1]

iv. closures $\lambda x C$ ('λ-abstractions').

(Of course, definitions of subconstructions, free/bound variables, open/closed constructions should be added here.) Recall that constructions are not λ-terms (which are expressions), λ-terms are only used to denote constructions. *Concepts* can be aptly explicated as certain constructions (e.g. Materna, 2004).

1.2 Simple theory of types

Early development of TIL was framed within the simple theory of types (*STT*; *cf.* Tichý, 1976)[2]. Let B (base) be a set of pair-wise disjoint collections of (primitive) objects:

a. Every member of B is a type over B.

b. If ξ, ξ_1, ..., ξ_n are (any) types over B, then $(\xi\, \xi_1 \ldots \xi_n)$, i.e. collection of total and partial functions from ξ_1, ..., ξ_n to ξ, is a type over B.

---

[1] Some compositions v-construct nothing.
[2] Of course, any STT is immune to Russell's paradox.

For the analysis of natural discourse Tichý utilized $B_{TIL} = \{\iota, o, \omega, \tau\}$ where ι collects *individuals*, o collects *truth-values* (just T and F), ω collects *possible worlds* (serving as a modal index), and τ collects *real numbers* (serving, *inter alia*, as a temporal index). Functions from possible world—moment of time couples are called *intensions*; these are *propositions, properties, relations-in-intension, individual offices*, etc. (Among non-intensions, the best known are classical unary or binary truth-functions, identity relation between ξ-objects, quantifiers as subclasses of classes of ξ-objects.)

1.3 Deduction

Tichý developed a deduction system with constructions (see papers in Tichý, 2004). Because of partiality, classical derivation rules are a bit modified, yet they are still rather classical. Derivation rules exhibit properties of (and relations between) objects and even certain properties of their constructions. I view definitions as certain $\Leftrightarrow$-rules.

1.4 Explication of meaning

In order to explicate meanings of (natural) language, Tichý employed a *semantic scheme* précised as follows:

an expression E
  | *expresses* (means) in L:
a construction   = the *meaning* of E in L
  | *constructs*:
an intension / non-intension = the *denotatum* of E in L

Empirical expressions ('the Pope', 'tiger', 'It rains in Nice', ...) denote intensions; non-empirical expressions ('not', '3', ...) denote non-intensions. The value of an intension in a possible world W at a time-moment T is the *referent*, in L, W and T, of an empirical expression. The denotatum (in L) and referent (in L, W, T) of a non-empirical expression are construed as identical.

To provide an example, the expression 'The Pope is popular' expresses the construction $\lambda w \lambda t [^0 \text{Popular}_{wt} {}^0\text{Pope}_{wt}]$. The construction constructs a proposition which maps world-time couples to T or F or nothing (truth-value gap). The proposition is the denotatum of the sentence in L. A particular value (if any) of the proposition in W, T is the referent of the sentence in L, W, T.

Well-known arguments show that intensional or 'sentencialistic' analyses of belief sentences (and other hyperintensional phenomena) are wrong. Tichý thus suggested to construe belief attitudes as attitudes towards constructions of propositions (not towards mere propositions or expressions): an agent only believes the construction expressed by the embedded sentence (and no other, though equivalent, construction). For instance, 'X believes that the Pope is popular' expresses the (2nd-order) construction

$$\lambda w \lambda t [{}^0\text{Believe}_{wt}\ {}^0X\ {}^0\lambda w \lambda t [{}^0\text{Popular}_{wt}\ {}^0\text{Pope}_{wt}]]$$

Note that ${}^0\lambda w \lambda t [{}^0\text{Popular}_{wt}\ {}^0\text{Pope}_{wt}]$ constructs just $\lambda w \lambda t [{}^0\text{Popular}_{wt}\ {}^0\text{Pope}_{wt}]$. Analogously, 'X calculates 3 ÷ 0' expresses $\lambda w \lambda t [{}^0\text{Calculate}_{wt}{}^0X\ {}^0[{}^03\ {}^0\div\ {}^00]]$; the agent is reported to have an attitude towards the procedure $[{}^03\ {}^0\div\ {}^00]$, not to its (non-existing) numerical result. Such explicit 'mentioning' of constructions by trivialization and other ways of constructing of constructions (e.g. via quantification over them) leads to the ramification of STT.

1.5 Tichý's type theory

For precise definition of Tichý's (ramified) type theory (TTT) see (Tichý, 1988, ch. 5). TTT has three layers:

1. STT (given above) which classifies first-order objects;

2. *1st-(2nd-, ..., n-)order constructions* (i.e. members of types $*_1, *_2, \ldots, *_n$, respectively) are constructions of 1st-(2nd-, ..., $n-1$-)order objects (or constructions);

3. functions from or to constructions (they belong, e.g., to the type $(*_1\tau)$).

The second level resembles to a Russellian ramified TT (RTT). Several kinds of cumulativity are inherent in TTT (e.g., every k-order construction is also a $(k+1)$-order construction). Known objections raised against Russell's RTT can be easily dismissed but one has to utilize a bit richer TTT than TTT over B_{TIL}.

I understand TTT as implementing four *Vicious Circle Principles* (*VCPs*).[3] Each of them is in fact a consequence of the *Principle*

---

[3] E.g. (Raclavský, 2009a).

of Specification: one cannot precisely specify an item by means of the item itself (already Russell stated such claim). The *Functional VCP*: no function can contain itself among its own arguments or values (*cf.* the layer 1.). The *Constructional VCP*: no construction can (v-)construct itself (*cf.* 2.; this VCP resembles to that of Russell); to illustrate, a variable c for constructions cannot be in its own range, it cannot v-construct itself—otherwise it would not be specifiable. The *Functional-Constructional VCP*: no function F can contain a construction of F among its own arguments or values (*cf.* 3.). The *Constructional-Functional VCP*: no construction C can (v-)construct a function having C among its own arguments or values (*cf.* 2. and 3.).

Concluding the section 1: unlike logical systems of rivalling solutions to SPs, it is explicitly stated what meanings are; the semantical theory is hyperintensional (not intensional or extensional), i.e. its underlying TT is ramified; the system is rather classical—bivalency and other classical logical laws are accepted, yet partiality is treated (thus logical laws are adapted).

2 Explicit approach

2.1 Language as hierarchy of codes

Language can be viewed as a normative system, such that people who conform to it are capable to exchange, communicate pieces of information. For our purposes it is sufficient to model language (in a synchronic sense) simply as a function from (Gödelized) expressions to meanings. Within TIL, a k-*order code* L^k is a function from real numbers to k-order constructions, it is an $(*_k\tau)$-object (Tichý, 1988, p. 228); there are various 1st-, 2nd-, ..., n-order codes (Tichý, 1988, p. 228).[4]

However, it is not sufficient to model (say) English by a single, say a 1st-order, code. Rather, a whole *hierarchy of codes* (called 'family' in Raclavský, 2009a) should be invoked as a model of English. The key reason consists in that English as a natural language is capable *to code*, to express by some of its expression, constructions of higher orders.

---

[4]Let me add that any grammatically correct composition of atomic expressions is included in a *sufficiently rich code*.

It has a connection with an important fact about codes. No construction of L^1, most notably $^0L^1$, is among constructions expressible-codable in L^1. Recalling the Functional-Constructional VCP, if $^0L^1$ would be a value of L^1, L^1 were not be specifiable at all. Unfortunately, $^0L^1$ is naturally understood as the meaning of 'L^1', the name of L^1. Thus when explicating '... in English ...' as expressive of $[\ldots {}^0L^1 \ldots]$, we need to take into account a higher-order code in which $[\ldots {}^0L^1 \ldots]$ is expressible.

From the just stated fact that *no construction of a k-order code L^k is codable in L^k* (only in a higher-order code) it follows that no expression referring to L^k is endowed with meaning in L^k (only in a higher-order code). By the compositionality-of-meaning principle, *no expression E, the subexpression of which refers to L^k, is endowed with meaning in L^k*.

Not any class of codes (of distinct orders) counts as a hierarchy of codes by which a particular language can be explicated. Some conditions should be imposed. A particular hierarchy of codes involves n codes $L^1, \ldots, L^n$ such that:

a. they are of n mutually distinct orders;

b. each expression having a meaning in L^k has the same meaning in L^{k+1};

c. an expression lacking meaning in L^k can be meaningful in L^{k+1}.

Of course, most of everyday communication takes place in the 1st-order code L^1 of a hierarchy. Higher-order coding means (e.g. L^2) of a hierarchy are invoked rarely—only when one comments parts of (say) English by means of the other parts of English (in this way I implement the universality-of-language principle).

Some remarks. Every code of the same hierarchy shares the same expressions (no predicates are forbidden); quantification over all of them is unrestricted. Due to the order-cumulativity of objects, every k-order code is also a $(k+1)$-order code, thus the type $(*_n\tau)$ includes (practically) all codes of the hierarchy; we can quantify over them. A hierarchy of codes is a certain class (it is an $(o(*_n\tau))$-object); thus one can quantify even over families. Finally, a hierarchy of codes is a 'system' of coding vehicles, not a particular vehicle ('language'); thus we investigate meanings of expressions in the members of a hierarchy, e.g. in L^n, not in the hierarchy as a whole.

2.2 Explication of semantic concepts

According to the 'explicit approach', semantic concepts (concepts of semantic properties and relations) are explicated as explicitly relative to language-code. Here are some sample definitions[5] (e.g. Raclavský, 2009a):

$[^0\text{TheMeaningOfIn}^n \, n \, l^n] \quad \Leftrightarrow^{*n} \quad [l^n \, n]$

$[^0\text{TheDenotatumOfIn}^\xi \, n \, l^n] \quad \Leftrightarrow^\xi \quad [^0\Gamma^{(\xi *_n)}[l^n \, n]]$

$[^0\text{TheReferentOfIn}^{I\zeta}_{wt} \, n \, l^n] \quad \Leftrightarrow^\zeta \quad [^0\Gamma^{(\xi *_n)}[l^n \, n]]_{wt}$

The construction $[l^n \, n]$ v-constructs the value (if any) of an n-order code L^n for the expression E, i.e. E's meaning in L^n. The function $\Gamma^{(\xi *_n)}$ maps any n-order construction C^n to the ξ-object (if any) v-constructed by C^n.

Truth can be construed as a property of propositions, constructions, and expressions (all defined in Raclavský, 2008). Truth as a *property of propositions* can be defined as follows, having thus 2 kinds of such properties (p ranges over propositions):

$[^0\text{True}^{\pi P}_{wt} p] \quad \Leftrightarrow^o \quad p_{wt}$

$[^0\text{True}^{\pi T}_{wt} p] \quad \Leftrightarrow^o \quad [^0\exists \lambda o[[o \, ^0= p_{wt}] \, ^0\wedge [o \, ^0= \, ^0T]]]$

The first defined concept is a concept in the *partial* ('P') sense: a proposition P can be neither true$^{\pi P}$ or false$^{\pi P}$; the latter is a concept in the *total* ('T') sense: a proposition P is true$^{\pi T}$ or not true$^{\pi T}$. Truth as a *property of constructions* have 4 kinds (each having n instances); a construction C^n is true^{*n} in W, T iff it v-constructs a proposition which is true$^\pi$ in W, T.

On the other hand, truth as a *property of expressions* is relative to a particular language-code (6 principal kinds):

$[^0\text{TrueIn}^P_{wt} \, n \, l^n] \quad \Leftrightarrow^o \quad [^0\text{True}^{\pi P}_{wt}[^0\Gamma^{(\pi *_n)}[l^n \, n]]]$

$[^0\text{TrueIn}^T_{wt} \, n \, l^n] \quad \Leftrightarrow^o \quad [^0\exists \lambda o[[o \, ^0= [^0\Gamma^{(\pi *_n)}[l^n \, n]]_{wt}] \, ^0\wedge [o \, ^0= \, ^0T]]]$

Note the interrelation of truth and the other basic semantic concepts: an expression E is true in L^n, W, T iff E expresses-means in L^n a construction of a proposition which is true$^\pi$ in W, T, i.e. E refers (in L^n, W, T) to T.

---

[5]Definitions can be aptly viewed as explications of the respective intuitive concepts.

Semantic Paradoxes and Transparent Intensional Logic

2.3 Solution to SPs

Let me illustrate the solution to particular SPs on the example of the (Belnap's) Paradox of Adder. Its paradox-producing expression is this:

D: '1 + the denotatum of D'

(The paradox: D denotes N; $N = 1 +$ the denotatum of D, i.e. $N = 1 + N$; but this is impossible because the adding-one function has no fixed point).

Here is my critical examination of the paradox:

a. If we do properly understand D, we have to bring out in which language-code the denotation of D proceeds.

b. One thus *disambiguates* D to (say) '1 + the denotatum of D in L^1' (hereafter simply D).

c. Thus our understanding of D takes place in the (say) 2nd-order code L^2 of English.

d. In L^2, D means the 2nd-order construction

$$[^0 1\ ^0 + [^0\text{TheDenotatumOfIn}_{wt}^\tau\ ^0\text{g}(D)\ ^0 L^1]]$$

(where $^0\text{g}(D)$ constructs the Gödelian number of D).

e. Being a 2nd-order construction, it cannot be expressed by D already in the 1st-order code L^1, thus D is without a meaning in L^1.

f. Lacking meaning in L^1, D has no denotatum in L^1.

g. The construction $[^0 1\ ^0 + [^0\text{TheDenotatumOfIn}_{wt}^\tau\ ^0\text{g}(D)\ ^0 L^1]]$ constructs nothing at all because the addition function obtains no suitable argument, since $[^0\text{TheDenotatumOfIn}_{wt}^\tau\ ^0\text{g}(D)\ ^0 L^1]$ constructs nothing.

h. The premise of the paradox, that D denotes a number N, is refuted.

Quite analogously for various Liars, e.g. S: 'S is not true'. The 2nd-order construction $\lambda w \lambda t [^0\neg [^0\text{TrueIn}^{\text{T}}_{wt}\, ^0g(S)\, ^0L^1]]$ is not expressible in L^1, but in L^2. In L^2, S denotes a false$^\pi$ proposition because there is no true$^\pi$ proposition denoted by S already in L^1. Hence I reject the premise of the respective paradox that the proposition denoted by S can be true$^\pi$.

Contingent or strengthened versions of SPs make no counter-examples for this kind of solution. All known principal paradoxes of denotation and reference are solved in (Raclavský, 2009a; Raclavský & Zouhar, 2011). All kinds of the Liar are solved in (Tichý, 1988, section 44; Raclavský, 2009b).

Such solution seems to be a certain mix of 'golden' ideas of Russell (VCP, hierarchy of propositional functions), Tarski (language/metalanguage) and perhaps also Kripke (partiality of a truth-predicate). Yet there are also significant dissimilarities. Unlike Russellian RTT, TTT treats both 'extensional' and 'intensional' functions; the latter ones, viz. constructions, are carefully individuated. Unlike in Tarski, language is explicated as a system of expressions coding meanings-constructions (which conform to the respective VCPs); moreover, semantic concepts are explicated as explicitly language-relative. Unlike in Kripke, semantic concepts in the total sense are explicated as well.

The important conclusion of the explicit approach: *semantic concepts-constructions involving a construction of a k-order code L^k are not expressible-codable in (sufficiently rich) L^k. Thus every code (of a hierarchy) is limited in its expressive power.*

3 Implicit approach

One may raise the following objection to the explicit approach. As a solution to SPs the explicit approach rightly applies only to those paradox-producing expressions in which language is explicitly referred to; however, typical paradox-producing expressions need no disambiguation to the form in which language is explicitly referred to; hence, a number of SPs remains in fact unresolved.

I can admit such objection. Nevertheless, I still claim that there is always at least *implicit* relativity to language of such semantic terms (and the terms are *ambiguous* after all).

In order to admit the objection, the following principle has to be adopted:

For every $(k+1)$-order construction of a property (relation) of expressions which involves a construction of a code L^k (that is l^k, $^0L^k$, etc.), there is an equivalent (v-congruent) k-order construction of the very same property (relation) involving no such construction of a code L^k.

To illustrate the principle, the 2nd-order construction:

$$\lambda w \lambda t \lambda n [^0\neg [^0\text{TrueIn}^T_{wt}\, n\, {}^0L^1]]$$

is equivalent to the 1st-order construction:

$$\lambda w \lambda t \lambda n [^0\neg [^0\text{True}^{TL^1}_{wt}\, n]]$$

Realize that $\lambda w \lambda t \lambda n [^0\neg [^0\text{True}^{TL^1}_{wt}\, n]]$ is *definable* by means of $\lambda w \lambda t \lambda n [^0\neg [^0\text{TrueIn}^T_{wt}\, n\, {}^0L^1]]$. Note also that '$L^1$' in '$^0\text{True}^{TL^1}$' indicates that the respective concept is related just to L^1, not to any other code (it is the definiens which shows that, i.e. removes the ambiguity of the respective intuitive concept). There is a number of such *implicitly language-relative semantic concepts*; my way of their explication is obvious.

Now, the expression 'not true' (without 'in') expresses in some code of the hierarchy the construction $\lambda w \lambda t \lambda n [^0\neg [^0\text{True}^{TL^1}_{wt}\, n]]$ (i.e. 'not true' is not disambiguated, e.g., to 'not true in L^1').

However, there is a danger of *revenge* of a paradox if one assumes that 'not true' expresses this construction already in the 1st-order code L^1.[6]

It is readily seen that the Functional-Constructional VCP and related principles are incapable to preclude the revenge (as they do in explicit cases). Thus I can appeal here to nothing but the *proof*— easily generalizable from Tichý's *Corollaries 44.1-4* (Tichý, 1988, pp. 292-293)—that a k-order code cannot code constructions like $\lambda w \lambda t \lambda n [^0\neg [^0\text{True}^{TL^k}_{wt}\, n]]$.

Here is the crucial idea of the proof. Assume, for *reductio*, that S expresses in L^1 a construction of a (total) proposition P, thus S denotes (in L^1) P; however, the construction $\lambda w \lambda t [^0\neg [^0\text{True}^{TL^1}_{wt}\, {}^0g(S)]]$ constructs a (total) proposition Q which is true$^\pi$ if the proposition

---

[6] Analogously for other semantic terms and concepts.

denoted by S in L^1 is not true$^\pi$ (Q is false$^\pi$ if the proposition denoted by S in L^1 is true$^\pi$); thus P cannot be identical with Q, hence S cannot express in L^1 a construction identical (or, more broadly, equivalent) with $\lambda w \lambda t [^0 \neg [^0 \text{True}_{wt}^{TL^1} {}^0 g(S)]]$.

How to explain this fact? As we have seen, concepts-constructions such as $\lambda w \lambda t \lambda n [^0 \neg [^0 \text{True}_{wt}^{TL^1} n]]$ are definable by means of constructions explicitly employing the code L^1. It follows that such concept is *relative to language-code after all*. Indeed, $\lambda w \lambda t \lambda n [^0 \neg [^0 \text{True}_{wt}^{TL^1} n]]$ and $\lambda w \lambda t \lambda n [^0 \neg [^0 \text{TrueIn}_{wt}^{T} n\, {}^0 L^1]]$ construct *one and the same property which is related to* L^1. (Hence, all semantic properties and relations are relative to language-code.) The purpose of any code is to discuss matters external to it; it is *not purpose of a code to discuss its own semantic features* (cf. Tichý, 1988, p. 231). We thus concluded, similarly as in the previous section, that *every code is limited in its expressive, coding power* (cf. Tichý, 1988, p. 233).

The final conclusion. Tarski's famous *Undefinability Theorem* says that semantic predicates concerning L are not definable in L. The TIL-approach to semantic concepts fully confirms it. Of course, it is added that the respective concepts are definable (the constructions exist and they may even construct something), yet they cannot be coded-expressed in a sufficiently rich L.[7]

References

Materna, P. (2004). *Conceptual systems*. Berlin: Logos.

Raclavský, J. (2008). Explikace druhů pravdivosti [Explications of kinds of being true]. *Sborník prací Filozofické fakulty brněnské univerzity, B* 53(1), 89–99.

Raclavský, J. (2009a). *Jména a deskripce: logicko-sémantická zkoumání* [Names and description: Logico-semantical investigations]. Olomouc: Nakladatelství Olomouc.

---

[7]A remark. A partial truth-predicate could be added only to that object language-code which has a *limited expressive power* (natural language is not such), i.e. a language not allowing to form a (meaningful) total *untruth*-predicate from the partial truth-predicate or a language not containing any equivalent of the total untruth-predicate. To illustrate the second possibility, consider $[^0 \text{Babig}_{wt} n] \Leftrightarrow^o [^0 \neg [^0 \exists \lambda o [[o\ ^0 = [^0 \text{TrueIn}_{wt}^{P} n\, ^0 L^1]]\ ^0 \wedge [o\ ^0 =\ ^0 T]]]]$ (the definiens is in fact a total concept of truth of expressions); one cannot safely add the predicate 'babig' so defined to the object-language L^1.

Raclavský, J. (2009b). Lhářský paradox, význam a pravdivost [Liar paradox, meaning and truth]. *Filosofický časopis*, 57(3), 325–351.

Raclavský, J., & Zouhar, M. (2011). *Paradoxes of denotations and reference*. (Manuscript)

Tichý, P. (1976). *Introduction to intensional logic*. (Unpublished manuscript)

Tichý, P. (1988). *The foundations of Frege's logic*. Walter de Gruyter.

Tichý, P. (2004). *Pavel Tichý's collected papers in logic and philosophy*. (V. Svoboda, B. Jespersen, & C. Cheyne, Eds.). University of Otago Press, Filosofia.

Jiří Raclavský
Department of Philosophy, Masaryk University
Arne Nováka 1, 602 00, Brno, Czech Republic
e-mail: raclavsky@phil.muni.cz
URL: http://www.phil.muni.cz/~raclavsky/

Unlimited Possibilities

Gonçalo Santos*

Abstract

I distinguish between a metaphysical and a logical reading of Generality Relativism. While the former denies the existence of an absolutely general domain, the latter denies the availability of such a domain. In this paper I argue for the logical thesis but remain neutral in what concerns metaphysics. To motivate Generality Relativism I defend a principle according to which a collection can always be understood as a set-like collection. I then consider a modal version of Generality Relativism and sketch how this version avoids certain revenge problems.

Keywords: absolute generality, Russell's paradox, plural quantification, indefinite extensibility, revenge

1 The incredulous stare

As many philosophers, when trying to explain the questions that occupy my mind, I often end up having to face that infamous look: the incredulous stare of my interlocutor. Personal experience has shown that if the goal is to get the stare in its most expressive format, a safe bet it to reply by saying that I'm worried about absolutely everything. This is almost always perceived as a joke, nevertheless it is quite true for I am in fact worried about everything. If these conversations didn't invariably change their subject, I would begin to provide a better characterization of my worry by following Rayo and Uzquiano (2006) in drawing a distinction between:

---

*The author wishes to thank to Philip Keller, Øystein Linnebo, José Martinez, Gabriel Uzquiano and Elia Zardini for manny helpful comments on previous versions of the present paper. This work was partially supported by the Centro de Filosofia da Universidade de Lisboa and by the FCT-project Hilbert's Legacy in the Philosophy of Mathematics, PTDC/FIL-FCI/109991/2009.

1. The metaphysical question: Is there an all inclusive domain of discourse?

2. The availability question: Is there an all inclusive domain of discourse available to us?

A domain that purports to be all inclusive, must accordingly include absolutely everything in it. To emphasize this even further, if there is an all inclusive domain, then absolutely nothing can lay outside of it. One way to understand the metaphysical question is thus as an inquiry concerning the extension of reality. Does it have limits of some kind? Is reality somehow open-ended or is it completely determined? Although I find these to be meaningful and deeply interesting issues, I will be here mostly concerned with the availability question. At first sight this might appear to be a less exciting topic, but I hope to convince the reader otherwise in the course of the paper. For as the metaphysical issue can be understood as being about the limits of reality, the availability question can be understood as an inquiry on the limits of language. The idea is that just as the alleged existence of an absolutely general domain sets a limit to what there is, the availability of such a domain of discourse would define the limits of what can be talked about.

It seems natural to ask if both questions are not closely intertwined. For instance, if someone denies the existence of an all inclusive domain, it seems that she will also have to deny that such a domain is available to us and contra-positively, if someone defends the availability of such a domain, it seems that she will also have to defend its existence. One suggestion to argue from the availability to the existence of an absolutely general domain would be the following. If there is an absolutely general domain of quantification which is available to us, that thing which is available must somehow exist. The argument would then need to show that the sense in which that thing exists is robust enough to imply an affirmative reply to the metaphysical question. One obvious way to resist the argument would be to deny that robustness and argue that the domain carries no metaphysical weight with it. Someone could perhaps claim that absolute generality is to be understood as a plurality. Since a plurality is to be understood as the mere sum of its elements there seems to be conceptual space for a position that argues for the availability of an absolutely general domain but remains free of a commitment with its metaphysical existence. This position will be considered in more detail further on in the paper, but for now I only wish to notice that the questions above are not intertwined in such a way that replying "yes" (or "no") to one of them, implies a "yes" (or accordingly, a "no") reply to the other. In particular, commitment with the availability of an absolutely general domain of quantification does not imply a commitment with the existence of an absolutely general domain. In the remanding of the paper, we shall use 'generality absolutist' to refer to someone who replies "yes" to the availability question and 'generality relativist' to someone who replies "no."

The plan then is the following. I assume that if an absolutely general domain of quantification is available, then it must be specifiable. That is, when we talk about absolutely everything, we should be able to specify the semantic value of 'absolutely everything' in an unequivocal manner. Presumably, that semantic value takes the form of a collection that somehow contains every object. If the proposed collection fails to contain every object that can be specified or even if its specification leaves open the possibility of there being objects that would not be part of it, the proposed collection cannot be legitimately taken to be absolutely general. I find this to be a reasonable assumption to make. In the absence of such a specification, the claim that an absolutely general domain of quantification is available sounds empty. I then argue that any collection purporting to be absolutely general can be extended. This sort of claim has been accused of being self-defeating but I will try to show how a modal formulation of generality relativism can be shown to avoid such a problem.

2 The availability question

A negative reply to the availability question might initially sound surprising. In fact, it is quite tempting to suppose that speakers can easily give a determinate specification of the domain of absolutely everything. For might they not say something like "The maximal domain is to consist of all objects?" When we say things like "Every object is self-identical" we indeed appear to be making an absolutely general claim. However, one reason to doubt this possibility is that the notion of 'object' appears not to lend itself to a determinate specification. For instance, a Meinongian who defends the existence of universes containing non-existent objects may be accused of incurring in conceptual or even factual mistakes. Nevertheless, it would be clearly unfair to accuse him of incompetence with the English language. In accounting for this situation, Glanzberg (2004, p. 549) says that the term 'object' is vague and as such, we are not able to decide among its different sharpenings just by appealing to its meaning.[1] The Meinongian, for instance, relies upon some sharpening of the term. The ordinary meaning of 'object' is however insufficient to decide whether that particular sharpening is preferable to another, more limited one. Moreover, and this is the crucial point, any attempt to sharpen its the meaning seems to run the risk of hitting upon a notion which is too narrow to really give us absolutely everything.

Glanzberg considers the following strategy to deal with the previous difficulty. Although there is no philosophical consensus concerning the status of meinongian non-existent objects, if we can understand the notion of *object* in a generous enough way to include all the metaphysical objects in the philosophical literature, that would give us an absolutely general domain of quantification. His idea is to specify something like the minimal conditions of objecthood, i.e., the conditions that need to be satisfied by something to qualify as an object. As Glanzberg puts it, equipped with such a definition we would be able to delineate the outer limits of 'object.' Although it might turn out that further philosophical investigation leads us to conclude that some of these logical objects have no metaphysical correlate, given a quantifier ranging over all of them, we could rest assured that no object with a metaphysical correlate falls outside of its range.

The task of providing a definition for the notion of 'object' might appear to be a non-starter. After all, this notion is so fundamental that it is hard to conceive of a *definiens* which does not presuppose it already. Still, here is an idea. In a well-known passage Russell (1937, p. 43) writes: "Whatever may be an object of thought, or may occur

---

[1] Notions like 'vague' and 'sharpening' have undoubtedly become terms of art in the philosophical literature. As such, Glanzberg's use of these notions in the present discussion might appear somewhat inadequate. For instance, vague terms lend themselves to *sorites* reasoning and there seems to be no reasoning of that kind associated with 'object'. Nevertheless, and although this appears to be a fair criticism, it does not carry much weight for the claim that Glanzberg is trying to make. His idea seems to be simply that 'object' can be said to be vagueness-like insofar as it shares an important feature of typically vague notions: it is by itself insufficient to decide among its potential sharpenings.

in any true or false proposition, or can be counted as one, I call a term. This, then, is the widest word in the philosophical vocabulary." We might then try to understand *logical object* as that to which a singular term refers. The point is not that every object has a name, but rather that in the present context having a name is sufficient to have an object. Accordingly, we might try to specify the domain of quantification of all the objects in the following way: the objects are the referents of singular terms. As we will now see, this does not serve as a specification for an absolutely general domain. The reason is that for any specification of the latter kind, we will be able to run a generalized version of Russell's paradox and name one object that could not belong to the specified domain.

Suppose that we have some specification of a domain. As a result, we can quantify over it. Hence, we can form the class term '$\{x: x = x\}$'. Let $y = \{x: x = x\}$. Then, by comprehension, there is $z = \{x \in y: x \notin x\}$. But $z \in z \leftrightarrow z \notin z$, so we get a contradiction.

In order to block this argument, the generality absolutist might first try to reject the step involving comprehension. Nevertheless, it is not obvious how to justify such a rejection. Generally, we can unequivocally determine whether an object is self-membered. Consequently, and given that the domain is supposed to be absolutely general, it is not immediately obvious why should the comprehension step fail to determine an object. On the other hand, if nothing is rejected in the previous argument the conclusion is clear. By using the most generous notion of 'object' available, we arrived at term that cannot refer to any object in the specified domain. Since having a term is sufficient for having an object, the argument gives us an object that must lie outside of the specified domain.

Glanzberg claims that the appeal to class abstracts, or sets does not play any role in the previous argument. According to him, assuming that y is a set or a proper class is not going to cause any substantial changes. It might however be worth considering whether appealing to pluralities might not help the generality absolutist. Perhaps that he will be able to block the Russellian argument above by arguing that the collection of absolutely everything is a plurality. We have briefly considered this idea before, when in the first section we discussed how the metaphysical and the availability question are independent from each other. The idea then was that by taking the collection of absolutely everything to be a plurality, one can talk about it even if the collection itself does not exist (in the sense that it does not carry any metaphysical weight). This was supposed to show that the metaphysical and the availability question are independent of each other. The question now becomes whether the generality absolutist could employ a similar move to block the previous Russellian argument. He could perhaps say that what the paradox reveals is that y fails to refer and that this is all that it takes to block the argument.

3 Strengthened Russell

I will now try to spell out the appeal to pluralities in more detail. This will at first sight offer a positive reply to the availability question. I will however end up arguing that the appeal to pluralities is not satisfactory because it seems to leave the generality absolutist begging the question.

Let us then begin by noticing that although the generality absolutist can deny that y (as used in the last section) refers, it follows from the availability of an absolutely general domain that we can talk about absolutely everything. In particular, we can still talk about the set of all the non-self-membered objects in it. That is, we seem to be able to define $z^* = \{\, x \prec y : x \notin x \,\}$. If we have thereby succeeded in defining an object, this needs to be part of absolutely everything and if that is the case, we do in fact end up with a new contradiction. For if z^* refers, we can certainly wonder whether z^* is self-membered and in the context of a proof, this inquiry takes the form of the biconditional $z^* \in z^* \leftrightarrow z^* \notin z^*$.

Perhaps that the generality absolutist can say that what this shows is that z^* also fails to refer. Maybe that he can say that z^* is identical to y. To do this, he only needs to assume that no object belongs to itself, a claim which does not really qualify as being controversial. So, if the absolutist makes this move, the closest we can get to a contradiction is $z^* \prec z^* \leftrightarrow z^* \notin z^*$. So far, so good for the absolutist then.

Let us now consider whether the generality relativist might still fight back at this point with some sort of strengthened Russell. Remember that in the Kripkean truth theory, when we reach the fixed point, the liar sentence cannot be found in the extension nor in the anti-extension of the truth predicate. Nevertheless and although we cannot express it in the language of the theory, it still seems correct to say that the liar sentence is not true. For a sentence to be true, it needs to be in the extension of the truth predicate. Given that the liar cannot be found there, it is not true. But since that is what the liar sentence itself says, we find ourselves back in contradiction. This is the so-called problem of the reinforced liar. Now, in the case at hand, although z^* has no elements, we can still talk about it. Thus, it still seems correct to say that z^* is not self-membered. We are now getting dangerously close to a new contradiction. The argument seems to stop here, however. For it to proceed, z^* would need to be a part of y but since z^* fails to refer, this is simply impossible.

4 Collapse

Appealing to pluralities thus appears to offer the generality absolutist a way out of the Russellian argument. I would however like to suggest that this is a deceiving appearance. To see this, begin by considering a principle which has been recently proposed by Linnebo (2010). In replying to the question "When do some things form a set?", Linnebo argues that nothing is required for some things to form a set. In other words, every plurality of objects collapses into a set. A bit more formally, say that some things xx *form a set* y just in case $\forall u\,(u \prec xx \leftrightarrow u \in y)$, and let $FORM(xx, y)$ abbreviate this claim.[2] Linnebo's principle can then be reformulated as,

(COLLAPSE) $\forall xx \exists y\, FORM(xx, y)$

Notice that a set is an entity over and above its elements. Let us then say that a collection is *set-like* if it is an object over and above its elements. Pluralities, on the other hand, distinguish themselves from sets in precisely this aspect. That is, a plurality is nothing over and above its elements. The principle that I would now like to propose parallels COLLAPSE in saying that all pluralities collapse into set-like collections. The only difference between the two principles is that mine is compatible with the existence of set-like collections other than sets. I will name this principle COLLAPSE*. The reason behind its extra ontological generosity will soon become clear.

---

[2]Notice that unlike second-order variables, plural variables range over the same objects as singular variables do.

I take it that one argument against COLLAPSE* would be the following. Assume that we can use plural quantifiers to talk about absolutely everything. If we collapse this collection into a set-like collection, there is an object that contains every object as an object. Thus, since the collection of absolutely everything contains absolutely everything, that collection would need to contain itself. But this cannot be.³ Therefore, COLLAPSE* fails in the absolutely general case.

This argument enjoys a certain plausibility. A set-like collection of absolutely everything arising out of COLLAPSE* would be an entity over and above its elements. But if we assume that we already have absolutely everything in the range of quantification, it would be impossible to bring more objects into it. Hence the failure of COLLAPSE*.

Notice however that the previous argument relies on a controversial assumption, namely, that we can quantify over absolutely everything. For the generality relativist this is the very issue under discussion. As such, he is free to accuse the previous argument of begging the question. Simultaneously, the relativist might highlight the plausibility of COLLAPSE*, by noticing some of the instances in which it obtains. If we have a plurality with any finite number of elements, for instance, nothing prevents us from having a set with precisely those elements. Likewise, pluralities with an infinite number of elements might also be collapsed into a set. One controversial case would be that of the collection of all sets. Assuming that there is a set of all sets leads to paradox. Nevertheless, nothing prevents us from collapsing that collection into a proper class. For our present purposes, the crucial thing to be noticed is that although they are not sets, proper classes are still set-like. The proper class that contains all sets is an object over and above its elements.

Our question now is, why should COLLAPSE* fail in the absolutely general case? Saying that this principle must fail because otherwise we would meet a contradiction is hardly an explanation. Contradictions do not have any explanatory power as they stand themselves in need of explanation. Therefore, without a principled reason for the failure of COLLAPSE*, the generality absolutist seems to be left in a question begging position. Appealing to pluralities cannot be his final word in this debate.

---

³Remember that to resist the generalized Russell, the absolutist needs to be committed with the thesis that there are no self-membered objects.

5 Revenge

We have seen how the generality relativist tries to use a version of Russell's paradox to motive a negative reply to the availability question. There is however a difficulty associated with this strategy. The claim that it leads to a self-defeating position can be found in several places in the literature. For instance, David Lewis (1991), Vann McGee (2000) and Timothy Williamson (2003), all put forward similar versions of an argument for this same claim. Essentially, their argument divides the relativist strategy into two stages. In the first stage, the relativist is supposed to come up with one counter-example to an hypothetical case of absolutely general quantification. In the second, the relativist generalizes the conclusion of the first stage, claiming that there is a counter-example for any hypothetical case of absolutely general quantification. A difficulty now arises in the form of a revenge problem. Notice that in order to put forward his last claim, the generality relativist needs to be committed with all the counter-examples. That is, if the relativist is to make a claim about all the hypothetical cases of absolutely general quantification, he needs nothing less than all the counter-examples. But then, how is it that making a claim about all these objects is not to be understood as an absolutely general claim? What other objects could there be? The generality relativist thus seems to contradict himself. While on the one hand he wants to deny absolutely general quantification, on the other he seems to need it in order to put forward his thesis.

In this section I will sketch how is it that Fine's (2006) modal formulation of generality relativism offers a way out of this difficulty. To do this, I begin by reconstructing the previous criticism within the framework proposed by Fine. He says that quantification is always relative to an interpretation and that the Russellian argument is capable of expanding any given interpretation. Generality relativism is then supposed to be understood as a thesis about interpretations of quantification. Following Fine, let $I, J, \ldots$ be variables for interpretations and $I_0, J_0, \ldots$ constants for particular interpretations. Read '$I \subset J$' as 'J *(properly) expands* I' and say that 'I *is extensible*' if possibly some interpretation extends it, i.e. $\Diamond \exists J\, (I \subset J)$. Generality relativism is then to be expressed in terms of two clauses:

(GR) $\forall I \Diamond \exists J\, (I \subset J)$

$(GR)^+ \quad \Box \forall I \Diamond \exists J \, (I \subset J)$

The revenge problem previously discussed can now be reformulated along the following lines. Consider (GR). This sentence makes a claim about all interpretations. In particular, it says that every interpretation can be expanded. Hence, its intended range of quantification cannot be anything less than all the interpretations. If (GR) leaves some interpretation out its range, the generality relativist would not be making a strong enough claim. In that situation, (GR) would leave open the possibility of there being an absolutely general interpretation outside of its range of quantification. Now, remember that according to Fine, quantification is always relative to some interpretation. Revenge takes place at this point. Since the generality relativist wants to make a claim about all interpretations, the interpretation associated with the quantifier in (GR) cannot be extended. But this reading of the sentence is not coherent with the claim that (GR) is supposed to be making. Namely, that all interpretations are extensible—in particular, the interpretation associated with the universal quantifier in (GR). Therefore, Fine's modal formulation of generality relativism seems to be committed with two irreconcilable claims. One of them presupposes quantification over all interpretation while the other implies that no quantifier ranges over all interpretations.

I now propose a possible way out of this problem. The basic idea is to put the modal vocabulary to work by making the domain of quantification change from world to world. Put differently, my suggestion is to interpret the formulas in our language by means of a variable domain semantics.[4] This semantics allows the range of quantification to change from world to word. In particular, (GR) is always interpreted as making a claim about all the interpretations in some possible world. That is, when we move into some world, (GR) makes a claim about the interpretations in that world and in that world only. Moreover, according to this semantics the range of quantification may vary between two worlds. Consequently, there might be interpretations in one world that do not exist in the other. That is, (GR) might talk about some interpretations in one world and about other interpreta-

---

[4] For a detailed presentation of the variable domain framework the reader is referred to (Hughes & Cresswell, 1996, p. 3, c. 5). I will restrict myself to an informal discussion of my proposed solution, but hope to present a more detailed account at some other point in the future.

tions in a different world. The idea then is the following. Assume that (GR) is asserted at world w_1. It says that all the interpretations in w_1 are extensible. We can unproblematically assume that none of these interpretations expands the interpretation associated with the universal quantifier in (GR). Eliminate this quantifier to obtain the formula,

$$\Diamond \exists J \, (I_0 \subset J) \qquad w_1$$

Notice that this formula says that extending I_0 implies a move into some possible world w_2. According to our semantics, we can assume that the range of quantification at w_2 is different from w_1. In particular, the interpretation associated with the universal quantifier in w_1 might exist in w_2. This semantics then allows us to say that the interpretation associated with the universal quantifier in w_1 can be expanded. We cannot say this in w_1 but nothing prevents us from doing so in w_2. Let us now move to w_2 and eliminate the existential quantifier to obtain,

$$(I_0 \subset J_0) \qquad w_2$$

Finally, notice that by appealing to (GR)$^+$ we can reproduce the same reasoning as before to show how any interpretation in w_2 can be extended. In a nutshell then, my suggestion is to use (GR) to talk about all the interpretations in some world and (GR)$^+$ to talk about all the worlds. Putting the two together, the generality relativist can talk about all interpretations in all worlds. In particular, he can say that they can all be extended.

6   Summing up

A complete account of the problem of absolute generality implies addressing many different issues. I have been here mainly concerned with one, the so-called *availability question*. This is the question of whether a domain of quantification where the variables range over absolutely everything is ever made available to us. I have argued that an appeal to plural quantification cannot be the generality absolutist final world on this issue. Moreover, I considered the problem of how could someone deny the availability of an absolutely general domain. The initial idea was to use a generalized version of Russell's paradox to do the job. Several authors have claimed that this strategy leads the generality relativist to an incoherent position. I then argued that interpreting Fine's modal version of generality relativism in terms of a variable domain semantics opens up a way around this difficulty.

References

Fine, K. (2006). Relatively unrestricted quantification. In A. Rayo & G. Uzquiano (Eds.), *Absolute generality* (pp. 20–44). Oxford: Oxford University Press.

Glanzberg, M. (2004). Quantification and realism. *Philosophy and Phenomenological Research*, *69*(3), 541–572.

Hughes, G. E., & Cresswell, M. J. (1996). *A new introduction to modal logic*. London and New York: Routledge.

Lewis, D. (1991). *Parts of classes*. Oxford: Blackwell Publishers.

Linnebo, Ø. (2010). Pluralities and sets. *The Journal of Philosophy*, *107*(3), 144–164.

McGee, V. (2000). Everything. In G. Sher & R. Tieszen (Eds.), *Between logic and intuition: Essays in honor of Charles Parsons*. Cambridge University Press.

Rayo, A., & Uzquiano, G. (Eds.). (2006). *Absolute generality*. Oxford: Oxford University Press.

Russell, B. (1937). *The principles of mathematics*. New York: W. W. Norton & Company, Inc.

Williamson, T. (2003). Everything. *Philosophical Perspectives*, *17*(1), 415–465.

Gonçalo Santos

LOGOS, Universitat de Barcelona,

C. Montalegre, 6–8, desp. 4023, Barcelona 08001

LanCog, Universidade de Lisboa,

Alameda da Universidade, Lisboa, 1600-214, Portugal

e-mail: `goncalo.b.santos@gmail.com`

Boxes Are Relevant

Igor Sedlár*

Abstract

We provide an embedding of the minimal positive relevant logic B$^+$ into a version of term-modal logic. We prove that $\alpha \in $ B$^+$ if and only if a translation of α into the term-modal language is valid in a particular class of term-modal models. The result demonstrates that relevant implication can be simulated in a modal framework.

Keywords: relevant logic, term-modal logic, translation

1 Introduction

As the familiar story goes, rendering "A implies B" as material implication $A \supset B$ or strict implication $\Box(A \supset B)$ leads to unfortunate consequences, to wit the notorious "paradoxes" of implication. Relevant logics were born of the need to set this right and to provide an appropriate rendering of "implies".

The common perception of the situation is that relevant logics *surpass* other logics in dealing with "implies". They add the necessary extra bit and prove $A \to B$ only if A is *relevant* to B. Hence, relevant implication is not subject to its own set of paradoxes. The justification of seeing things this way seems to be the above-mentioned existence of "paradoxes" of other kinds of implication. For example, relevant logics surpass normal modal logics, since every normal modal logic proves $\Box(\Box A \supset \Box(B \supset A))$. However, if we see $\Box(A \supset B)$ as a rendering

---

*This paper was written in the Department of Logic and Methodology of Science, Comenius University, as a part of the research project VEGA 1/0046/11, *Semantic models, their explanatory power and applications*. The audience at *Logica 2011* deserves my gratitude for valuable comments. I have especially benefited from discussions with David Makinson, Ed Mares, Jarda Peregrin, Adam Přenosil, and Sebastian Sequoiah-Grayson.

of "A implies B", then this formula may be read as "The fact that A is necessarily true implies that A is implied by any B whatsoever". Surely, a not so plausible claim about the nature of "implies". But, one might ask, is this fact a good enough reason to assume that there is *no* relevant rendering of "A implies B" in a modal language?

This issue has been partly addressed by R. K. Meyer as the *Lewis problem*, see (Meyer, 1975). Meyer argued that one cannot successfully simulate relevant implication by means of formulas of the monomodal language. However, modal logic has evolved quite a bit since the times of C. I. Lewis and the monomodal language is only the tip of an extensive iceberg. The zoo of modal languages includes multimodal languages, languages with polyadic modalities, dynamic languages and what not.

This paper demonstrates that, with a specific relevant system in mind, relevant implication *can* be simulated by means of a modal language. To be more specific, we embed the positive basic relevant logic B$^+$ into an adjusted version of *term-modal logic*. Term-modal logics are extensions of first-order multimodal logics that allow to name modal operators by arbitrary terms of the language.

Interestingly, our approach has been foreshadowed in the relevant logic community. In the 1970's, Peter Woodruff suggested to treat the Routley-Meyer frames with the ternary accessibility relation R as multimodal frames with a family of *binary* relations R_x indexed by points of the frame.[1] This paper might be perceived as an exploration of Woodruff's idea.[2]

The paper is organised as follows. Section 2 sketches the basic features of B$^+$. We find the simplified semantics for B$^+$ particularly suitable for our purposes. In Section 3, we use the *standard translation* technique, customary in modal logic, to provide a first-order analogue of relevant formulas. We argue that the possibility to apply standard translation to relevant formulas demonstrates that relevant implication can be simulated by classical first-order logic. Section 4 provides an outline of term-modal logics and specifies the version of these logics we shall make use of. We prove our main result in Section 5. Section 6 concludes the paper and sketches some possible directions of future research.

---

[1] See (Dunn & Meyer, 1997) and (Beall et al., in press), for example.

[2] However, the underlying idea of this paper was conceived of by the author prior to getting acquainted with Woodruff's suggestion.

2 Simplified semantics for B⁺

This section defines the basic syntactic and semantic notions related to B⁺. We use the simplified semantics of (Priest & Sylvan, 1992).

Definition 1 The *positive relevant language* $\mathcal{L}_r$ contains a denumerable set $\Phi = \{p_1, p_2, \ldots\}$ of propositional variables and binary connectives $\wedge, \vee, \rightarrow$. The set of formulas of $\mathcal{L}_r$, Fm_r, is defined in the usual fashion:
$$\alpha ::= p \mid \alpha \wedge \alpha \mid \alpha \vee \alpha \mid \alpha \rightarrow \alpha$$

Definition 2 A *simplified relevant model* is a structure
$$\mathbf{M} = (K, g, R, I)$$

where K is a non-empty set (of "points", where $w, w', \ldots$ are variables ranging over points), $g \in K$, $R \subseteq K^3$ such that $Rgww'$ iff $w = w'$ and I is a function that assigns to every (α, w) a member of $\{0, 1\}$. The function behaves as follows:

1. $I(\alpha \wedge \beta, w) = 1$ iff $I(\alpha, w) = I(\beta, w) = 1$

2. $I(\alpha \vee \beta, w) = 1$ iff $I(\alpha, w) = 1$ or $I(\beta, w) = 1$

3. $I(\alpha \rightarrow \beta, w) = 1$ iff for all $w'w''$, if $Rww'w''$ and $I(\alpha, w') = 1$, then $I(\beta, w'') = 1$

A formula α is *valid in* $\mathbf{M}$ ($\mathbf{M} \models \alpha$) iff $I(\alpha, g) = 1$. B⁺ is the set of formulas α such that $\mathbf{M} \models \alpha$ for every $\mathbf{M}$.

3 The standard translation

In modal logic, the standard translation is a function ST_x that transforms a modal formula φ into a classical first-order formula $ST_x\varphi$ which states the truth-conditions of φ at x.[3]

Interestingly, the same technique may be used in the context of Fm_r to obtain first-order equivalents of relevant formulas.[4]

---

[3] See (Blackburn, Rijke, & Venema, 2001), for example.
[4] A similar translation for the Lambek Calculus was provided in (Kurtonina, 1995). See also (Kurtonina, 1998).

Definition 3 The *classical correspondence language* $\mathcal{L}_c$ is a first-order language with the signature consisting of denumerably many unary predicates $P_1, P_2, \ldots$ (corresponding to $p_1, p_2, \ldots$), a constant a (corresponding to g) and a ternary relation symbol T.

Definition 4 Let x be a first-order variable. The *standard translation* is a function ST_x that takes members of Fm_r to formulas of $\mathcal{L}_c$ such that:

1. $ST_x(p_i) = P_i x$

2. $ST_x(\alpha \wedge \beta) = ST_x(\alpha) \wedge ST_x(\beta)$

3. $ST_x(\alpha \vee \beta) = ST_x(\alpha) \vee ST_x(\beta)$

4. $ST_x(\alpha \to \beta) = \forall yz((Txyz \wedge ST_y(\alpha)) \supset ST_z(\beta))$

where the variables y, z are new.

For example, $ST_x(p \to (q \to p))$ is calculated as follows:

$$\begin{aligned}ST_x(p \to (q \to p)) &= \forall yz((Txyz \wedge ST_y(p)) \supset ST_z(q \to p)) \\ &= \forall yz((Txyz \wedge Py) \supset \forall y'z'((Tzy'z' \wedge ST_{y'}(q)) \\ &\quad \supset ST_{z'}(p)) \\ &= \forall yz((Txyz \wedge Py) \supset \forall y'z'((Tzy'z' \wedge Qy') \supset Pz'))\end{aligned}$$

Now, simplified relevant models may be seen as first-order models for $\mathcal{L}_c$. To be more specific, we may see $\mathbf{M}$ as a first-order model $M^{\mathbf{M}} = (K, Int)$, where Int is an interpretation function such that

1. $w \in Int(P_i)$ iff $I(p, w) = 1$

2. $Int(a) = g$

3. $(w, w', w'') \in Int(T)$ iff $Rww'w''$

Let $ST_a(\alpha)$ be $ST_x(\alpha)$ with every free occurrence of x replaced by a. Moreover, let $M^{\mathbf{M}} \models ST_x(\alpha)[x := w]$ mean that $ST_x(\alpha)$ holds in $M^{\mathbf{M}}$ under any valuation V such that $V(x) = w$. The following lemma shows that considering first-order translations of relevant formulas instead of the relevant formulas themselves is appropriate.

Lemma 1

a) For all $w \in K$, $I(\alpha, w) = 1$ iff $M^\mathbf{M} \models ST_x(\alpha)[x := w]$

b) $\mathbf{M} \models \alpha$ iff $M^\mathbf{M} \models ST_a(\alpha)$

c) $\mathbf{M} \models \alpha \to \beta$ iff $M^\mathbf{M} \models \forall x\, (ST_x(\alpha) \supset ST_x(\beta))$

Proof. a) The claim holds trivially for $p_i, \wedge, \vee$. Assume that $I(\alpha \to \beta, w) = 1$. This means that for all $w'w''$, if $Rww'w''$ and $I(\alpha, w') = 1$, then $I(\beta, w'') = 1$. Hence, for all yz such that $Twyz$, if $M^\mathbf{M} \models ST_y(\alpha)$, then $M^\mathbf{M} \models ST_z(\beta)$. But this holds iff $M^\mathbf{M} \models ST_x(\alpha \to \beta)[x := w]$.

b) This claim is a trivial consequence of a)

c) This claim follows from the above claims and the fact that $Rgww'$ iff $w = w'$, stated in definition 2. □

Lemma 1 may be seen as showing that relevant implication can be simulated by classical first-order logic. This is rather trivial, however, since $ST_x(\alpha)$ mimics the truth-conditions of α at w, if $V(x) = w$ for a first-order valuation V. Nevertheless, one may devise a nice philosophical interpretation of the situation. K may be seen as a group of agents, where g is the "leader" of the group. $P_i x$ corresponds to the claim that agent x has the information that p_i holds. Obviously, x has the information that $A \wedge B$ iff x has the information that A and also that B. The clause for disjunction suggests that agents behave "intuitionistically": x has the information that $A \vee B$ iff x has the information that A or has the information that B. The relation T (corresponding to R) may be seen as a ternary relation of "supposed information source": $Txyz$ means that the agent x thinks that y is an information source for z. This fits in nicely with the truth-condition for relevant implication. Agent x has the information that $A \to B$ iff every z, such that x thinks that a y is an information source for z and y in fact has the information that A, has the information that B. In general, this invites one to see B$^+$ as a first-order theory with $\forall yz(Tayz \leftrightarrow y = z)$ as the sole non-logical axiom.

In Section 5 we show how to "modalise" this theory. We shall prove that relevant implication can be simulated by term-modal logics. But first, we have to explain what term-modal logics are.

4 Term-modal logics

Term-modal logics are extensions of first-order modal logic where naming of modalities by terms of the language is permitted, (Fitting, Thalmann, & Voronkov, 2001; Thalmann, 2000). (Here we shall only sketch the basics of standard term-modal logics and we shall move quickly to our adjusted logics.) The primary interpretation of the labelled modalities is epistemic. An example of a term-modal formula is $\exists x[x]Px$, stating that there is an agent that knows that *she* has property P. Another example might be $\forall x(Px \supset [x]Px)$ stating that every agent with property P knows that she has the property P. Obviously, term-modal logics are rather expressive and interesting.

In general, the central semantic structures related to term-modal logics are *frames over* $\mathbf{D}$, where $\mathbf{D}$ is a non-empty set, the *domain*. The domain provides the universe of quantification. Frames over $\mathbf{D}$ contain a set of points W and a ternary relation, which is a subset of $W \times \mathbf{D} \times W$. Obviously, the ternary relation may be seen as a collection of binary relations labelled by members of $\mathbf{D}$.

We shall work with an adjusted version of the standard term-modal logics, which we call *propositional term-modal logic*. The tag "propositional" is not to be taken at face value, however, since our language is first-order after all. To motivate the tag, observe that a connection between the relevant ternary relation and the term-modal ternary relation is readily provided by setting $\mathbf{D} = W$. This shall allow us to use the machinery of term-modal logics to quantify over points. Moreover, the language of our logic shall contain an unlabelled modality $\square$.

Definition 5 The *propositional term-modal language* $\mathcal{L}_t$ is an extension of the first-order modal language. Hence, it contains a signature Σ. The signature Σ consists of denumerably many unary predicates $P_1, P_2, \ldots$, a binary relation symbol S and a constant a. Let *Var* be a denumerable set of variables $x_1, x_2, \ldots$. Every constant and variable is a term of the language $\mathcal{L}_t$. The set of formulas of $\mathcal{L}_t$ is defined as follows:

1. If t, t' are terms, then $P_i t$ and Stt' are atomic formulas, for every i.

2. If φ is a formula (whether atomic or not), then $\square \varphi$, $\neg \varphi$, $\varphi \wedge \psi$, $\varphi \vee \psi$ and $\varphi \supset \psi$ are formulas.

3. If t is a term and φ is a formula, then $[t]\varphi$ is a formula

4. If x is a variable and φ is a formula, then $\forall x \varphi$ is a formula.

(Note that our term-modal language is not positive.)

The central semantic structures of propositional term-modal logics are propositional frames.

Definition 6 A *propositional frame* is a structure $\mathfrak{F} = (W, \mathbf{R}, \longrightarrow)$, where W is a nonempty set (of points, where $\mathbf{x_1}, \mathbf{x_2}, \ldots$ are variables ranging over points), $\mathbf{R} \subseteq W^2$ and $\longrightarrow \subseteq W^3$. $(\mathbf{x}, \mathbf{y}, \mathbf{z}) \in \longrightarrow$ shall be written as $\mathbf{x} \xrightarrow{\mathbf{y}} \mathbf{z}$.

Note that propositional frames are a special case of the term-modal frames over $\mathbf{D}$ where $\mathbf{D} = W$. Moreover, they contain an unlabelled binary relation $\mathbf{R}$.

To be able to state truth-conditions of formulas at points, we have to make use of an interpretation function that assigns to every point and member of Σ an appropriate value of the member of Σ at the point.

Definition 7 A *propositional model* is a structure $\mathfrak{M} = (\mathfrak{F}, \mathfrak{I})$, where $\mathfrak{F}$ is a propositional frame and $\mathfrak{I}$ is an interpretation function such that:

1. $\mathfrak{I}(\mathbf{x}, a) \in W$ and $\mathfrak{I}(\mathbf{x}, a) = \mathfrak{I}(\mathbf{y}, a)$, for all $\mathbf{x}, \mathbf{y} \in W$

2. $\mathfrak{I}(\mathbf{x}, P_i) \subseteq W$ and $\mathfrak{I}(\mathbf{x}, P_i) = \mathfrak{I}(\mathbf{y}, P_i)$, for all $\mathbf{x}, \mathbf{y} \in W$ and every i

3. $\mathfrak{I}(\mathbf{x}, S) \subseteq W^2$

A valuation V is a function from Var to W. V' is a x-variant of x iff $V(y) = V'(y)$ for all y such that $x \neq y$. The truth-conditions of formulas in points with respect to valuations are given as follows:

1. $\mathfrak{M}, \mathbf{x}, V \Vdash P_i a$ iff $\mathfrak{I}(\mathbf{x}, a) \in \mathfrak{I}(\mathbf{x}, P_i)$

2. $\mathfrak{M}, \mathbf{x}, V \Vdash P_i x$ iff $V(x) \in \mathfrak{I}(\mathbf{x}, P_i)$

3. $\mathfrak{M}, \mathbf{x}, V \Vdash Sxy$ iff $(V(x), V(y)) \in \mathfrak{I}(\mathbf{x}, S)$ (similarly for Sax and Sxa)

4. The truth-conditions for boolean combinations of formulas are as usual

5. $\mathfrak{M}, \mathbf{x}, V \Vdash \Box\varphi$ iff $\mathfrak{M}, \mathbf{y}, V \Vdash \varphi$, for all $\mathbf{y}$ such that $\mathbf{Rxy}$

6. $\mathfrak{M}, \mathbf{x}, V \Vdash [a]\varphi$ iff for all $\mathbf{y}$ such that $\mathbf{x} \xrightarrow{\mathfrak{I}(\mathbf{x},a)} \mathbf{y}$, $\mathfrak{M}, \mathbf{y}, V \Vdash \varphi$

7. $\mathfrak{M}, \mathbf{x}, V \Vdash [x]\varphi$ iff for all $\mathbf{y}$ such that $\mathbf{x} \xrightarrow{V(x)} \mathbf{y}$, $\mathfrak{M}, \mathbf{y}, V \Vdash \varphi$

8. $\mathfrak{M}, \mathbf{x}, V \Vdash \forall x \varphi$ iff for all x-variants V' of V, $\mathfrak{M}, \mathbf{x}, V' \Vdash \varphi$

φ is valid in $\mathfrak{M}$ iff $\mathfrak{M}, \mathbf{x}, V \Vdash \varphi$, for every $\mathbf{x}, V$. φ is valid in a class of models iff it is valid in every model in the class.

Note that in propositional models, the interpretation of a and P_i is constant over W, for every i. Thus, the predicates P_i behave like propositional variables. This is another explanation of our use of the term "propositional". Moreover, this enables us to simplify notation and write $\mathfrak{I}(a)$ instead of $\mathfrak{I}(\mathbf{x}, a)$ and $\mathfrak{I}(P_i)$ instead of $\mathfrak{I}(\mathbf{x}, P_i)$. Note also that S is a binary version of the ternary T from Section 3: If $V(x) = \mathbf{x}$ and $V(y) = \mathbf{y}$, then the fact that Sxy holds at $\mathbf{z}$ under V may be perceived to mean that $\mathbf{z}$ "thinks" that $\mathbf{y}$ is accessible from $\mathbf{x}$.

Despite the fact that propositional models are rather specific, we shall concentrate on an even more specific subclass of them.

Definition 8 A *specific propositional model* is a model such that

1. $\mathbf{y} \xrightarrow{\mathbf{x}} \mathbf{z}$ iff $(\mathbf{y}, \mathbf{z}) \in \mathfrak{I}(\mathbf{x}, S)$

2. $\mathbf{Rxy}$ iff $\mathbf{y} \xrightarrow{\mathbf{x}} \mathbf{z}$ for some $\mathbf{z}$

3. $(\mathbf{y}, \mathbf{z}) \in \mathfrak{I}(\mathfrak{I}(a), S)$ iff $\mathbf{y} = \mathbf{z}$

In specific models, the interpretation of the binary relation symbol S is always in line with $\longrightarrow$ (clause 1.). The relation $\mathbf{R}$ represents "one half" of $\longrightarrow$ (clause 2.). Moreover, $\mathbf{y} \xrightarrow{\mathfrak{I}(a)} \mathbf{z}$ iff $\mathbf{y} = \mathbf{z}$ (clause 3.). Thus, $\mathbf{R}\mathfrak{I}(a)\mathbf{y}$ for all $\mathbf{y} \in W$. It is readily seen that specific models mimic simplified relevant models.

5 The main result

This section contains the proof of our main result to the effect that $\alpha \in \mathsf{B}^+$ iff a translation of α is valid in the class of specific propositional models. First, we define two translation functions. Then we shall prove that for every $\alpha \notin \mathsf{B}^+$, the translation of α has a specific propositional countermodel. Afterwards we shall prove the converse, i.e. that if the translation of α has a specific propositional countermodel, then $\alpha \notin \mathsf{B}^+$.

5.1 The translation functions

Definition 9 Let $x \in Var$. The *term-modal translation* ST'_x is a function from Fm_r to formulas of $\mathcal{L}_t$ such that:

1. $ST'_x(p_i) = P_i x$

2. $ST'_x(\alpha \wedge \beta) = ST'_x(\alpha) \wedge ST'_x(\beta)$

3. $ST'_x(\alpha \vee \beta) = ST'_x(\alpha) \vee ST'_x(\beta)$

4. $ST'_x(\alpha \to \beta) = \forall yz(Syz \supset \Box(ST'_y(\alpha) \supset [x]ST'_z(\beta)))$

Definition 10 The *a-translation* ST'_a is a function from Fm_r to formulas of $\mathcal{L}_t$ such that:

1. $ST'_a(p_i) = P_i a$

2. $ST'_a(\alpha \wedge \beta) = ST'_a(\alpha) \wedge ST'_a(\beta)$

3. $ST'_a(\alpha \vee \beta) = ST'_a(\alpha) \vee ST'_a(\beta)$

4. $ST'_a(\alpha \to \beta) = \Box \forall x (ST'_x(\alpha) \supset ST'_x(\beta))$

As an example, we give translations of $p \to (q \to p)$:

$$ST'_x(p \to (q \to p)) = \forall yz(Syz \supset \Box(ST'_y(p) \supset [x]ST'_z(q \to p)))$$
$$= \forall yz(Syz \supset \Box(Py \supset [x]\forall y'z'(Sy'z'$$
$$\supset \Box(ST'_{y'}(q) \supset [z]ST'_{z'}(p)))))$$
$$= \forall yz(Syz \supset \Box(Py \supset [x]\forall y'z'(Sy'z'$$
$$\supset \Box(Qy' \supset [z]Pz'))))$$
$$ST'_a(p \to (q \to p)) = \Box\forall x(ST'_x(p) \supset ST'_x(q \to p)))$$
$$= \Box\forall x(Px \supset \forall yz(Syz \supset \Box(ST'_y(q) \supset [x]ST'_z(p))))$$
$$= \Box\forall x(Px \supset \forall yz(Syz \supset \Box(Qy \supset [x]Pz)))$$

5.2 Modal twins

Definition 11 A *modal twin* of a simplified relevant model **M** is a propositional model $\mathfrak{M}^{\mathbf{M}} = (W, \mathbf{R}, \longrightarrow, \mathfrak{I})$ such that:

1. $W = K$

2. $\mathfrak{I}(a) = g$

3. $w' \xrightarrow{w} w''$ iff $Rww'w''$

4. $w \in \mathfrak{I}(P_i)$ iff $I(p_i, w) = 1$

Lemma 2 $\mathfrak{M}^{\mathbf{M}}$ *is a specific simple model, for every* **M**.

Proof. Follows immediately from the definitions. □

Lemma 3 *Let* $\mathbf{M} = (K, g, R, I)$ *be a simplified relevant model and* $\mathfrak{M}^{\mathbf{M}}$ *be its modal twin. For all* $w \in K$, $I(\alpha, w) = 1$ *iff* $\mathfrak{M}^{\mathbf{M}}, w, V \Vdash ST'_x(\alpha)[x := w]$.

Proof. The proof is trivial for p, $\wedge$, $\vee$. To prove the claim for $\alpha \to \beta$, assume first that $\mathfrak{M}^{\mathbf{M}}, w, V \nVdash ST'_x(\alpha \to \beta)$ for some point w and some V such that $V(x) = w$. By the definition of ST', this means that $\mathfrak{M}^{\mathbf{M}}, w, V \Vdash \exists yz \left(Syz \wedge \Diamond(ST'_y(\alpha) \wedge \langle x \rangle \neg ST'_z(\beta)) \right)$. Hence, for some (y, z)-variant V' of V, $\mathfrak{M}^{\mathbf{M}}, w, V' \Vdash Syz \wedge \Diamond(ST'_y(\alpha) \wedge \langle x \rangle \neg ST'_z(\beta))$. By the definition of specific models, $\mathbf{R}wV'(y)$. Hence, $\mathfrak{M}^{\mathbf{M}}, V'(y), V' \Vdash ST'_y(\alpha)$ and $\mathfrak{M}^{\mathbf{M}}, V'(y), V' \Vdash \langle x \rangle \neg ST'_z(\beta)$. Since $V'(x) = w$, we have $V'(y) \xrightarrow{w} V'(z)$ and $\mathfrak{M}^{\mathbf{M}}, V'(z), V' \nVdash ST'_z(\beta)$. If $V'(y) = w'$ and $V'(z) = w''$, then, by the definition of a modal twin, $Rww'w''$ in the original relevant model. By the induction hypothesis, $I(\alpha, w') = 1$ and $I(\beta, w'') = 0$. Thus, $I(\alpha \to \beta, w) = 0$.

Conversely, assume that $I(\alpha \to \beta, w) = 0$ for some w. This means that we have w', w'' such that $Rww'w''$, $I(\alpha, w') = 1$ and $I(\beta, w'') = 0$. By the induction hypothesis, $\mathfrak{M}^{\mathbf{M}}, w', V' \Vdash ST'_x(\alpha)$ if $V'(x) = w'$, and $\mathfrak{M}^{\mathbf{M}}, w'', V'' \Vdash \neg ST'_x(\beta)$ if $V''(x) = w''$. Now rename the variables (bound and free) so that $ST'_x(\alpha)$ becomes $ST'_y(\alpha)$ and $ST'_x(\beta)$ becomes $ST'_z(\beta)$.[5] Let V be any valuation such that $V(x) = w$, $V(y) = V'(x)$ and $V(z) = V''(x)$. Obviously, $V(y) \xrightarrow{w} V(z)$. Hence, $\mathfrak{M}^{\mathbf{M}}, w, V \Vdash Syz \wedge \Diamond(ST'_y(\alpha) \wedge \langle x \rangle \neg ST'_z(\beta))$. Thus, $\mathfrak{M}^{\mathbf{M}}, w, V \Vdash \exists yz(Syz \wedge \Diamond(ST'_y(\alpha) \wedge \langle x \rangle \neg ST'_z(\beta)))$. In other words, $\mathfrak{M}^{\mathbf{M}}, w, V \nVdash ST'_x(\alpha \to \beta)$. □

Lemma 4 *For all* $\mathbf{M}$ *and* $\mathfrak{M}^{\mathbf{M}}$, $I(\alpha, g) = 1$ *iff* $\mathfrak{M}^{\mathbf{M}}, \mathfrak{I}(a) \Vdash ST'_a(\alpha)$.

Proof. The lemma holds trivially for p, $\wedge$, $\vee$. To prove the claim for $\alpha \to \beta$, assume that $\mathfrak{M}^{\mathbf{M}}, \mathfrak{I}(a) \nVdash ST'_a(\alpha \to \beta)$. Since $ST'_a(\alpha \to \beta)$ contains no free occurrences of variables, this holds iff $\mathfrak{M}^{\mathbf{M}}, \mathfrak{I}(a), V \Vdash \Diamond \exists x (ST'_x(\alpha) \wedge \neg ST'_x(\beta))$ for all V. Since $\mathbf{R}\mathfrak{I}(a)w$ for all $w \in W$, the above claim holds iff there is a $w \in W$ such that $\mathfrak{M}^{\mathbf{M}}, w, V \Vdash \exists x(ST'_x(\alpha) \wedge \neg ST'_x(\beta))$ for all V. By the definition of $\mathfrak{M}^{\mathbf{M}}$ and Lemma 3, this is the case iff there is a $w \in K$ such that $I(\alpha, w) = 1$ and $I(\beta, w) = 0$. Hence, iff $I(\alpha \to \beta, g) = 0$. □

Lemma 5 *If* $\alpha \notin \mathsf{B}^+$, *then* $ST'_a(\alpha)$ *is not valid in the class of specific models.*

Proof. The Lemma follows immediately from Lemmas 2 and 4. □

5.3 Relevant twins

Definition 12 A *relevant twin* of a specific model $\mathfrak{M}$ is a model $\mathbf{M}^{\mathfrak{M}} = (K, g, R, I)$ such that

1. $K = W$
2. $g = \mathfrak{I}(a)$
3. $Rww'w''$ iff $w' \xrightarrow{w} w''$

---

[5] For example, if $\alpha = p \wedge q$, then $ST'_x(\alpha) = Px \wedge Qx$. After renaming, this turns into $Py \wedge Qy$.

4. $I(p_i, w) = 1$ iff $w \in \mathfrak{I}(P_i)$, for $w \neq g$

5. $I(p_i, g) = 1$ iff $\mathfrak{I}(a) \in \mathfrak{I}(P_i)$

Lemma 6 $\mathbf{M}^{\mathfrak{M}}$ *is a simplified relevant model, for every $\mathfrak{M}$.*

Proof. Immediate from the definitions. □

Lemma 7 *For all $w \in W$, $\mathfrak{M}, w, V \Vdash ST'_x(\alpha)[x := w]$ iff $I(\alpha, w) = 1$ in $\mathbf{M}^{\mathfrak{M}}$.*

Proof. Again, the proof is trivial for p, $\wedge$, $\vee$. To prove the claim for $\alpha \to \beta$, assume that $\mathfrak{M}, \mathbf{x}, V \Vdash ST'_x(\alpha \to \beta)$, where $V(x) = \mathbf{x}$. This means that $\mathfrak{M}, \mathbf{x}, V \Vdash \forall yz(Syz \supset \Box(ST'_y(\alpha) \supset [x]ST'_z(\beta)))$. However, $\mathfrak{M}, \mathbf{x}, V' \Vdash Syz$ iff $V'(y) \xrightarrow{\mathbf{x}} V'(z)$, for every V'. Thus, the above claim amounts to: If $\mathbf{y} \xrightarrow{\mathbf{x}} \mathbf{z}$ and $\mathfrak{M}, \mathbf{y}, V \Vdash ST'_y(\alpha)$, then $\mathfrak{M}, \mathbf{z}, V \Vdash ST'_z(\beta)$. By the definition of $\mathbf{M}^{\mathfrak{M}}$ and the induction hypothesis, if $V(x) = w$, then for all $w'w''$ such that $Rww'w''$, if $I(\alpha, w') = 1$, then $I(\beta, w'') = 1$. Thus, $I(\alpha \to \beta, w) = 1$. □

Lemma 8 *Let $\mathfrak{M}$ be a specific model and $\mathbf{x}$ a point in the model. For every V, if $\mathfrak{M}, \mathbf{x}, V \nVdash ST'_a(\alpha)$, then $\mathbf{M}^{\mathfrak{M}} \nvDash \alpha$.*

Proof. Assume that $\mathfrak{M}, \mathbf{x}, V \nVdash ST'_a(p)$, i.e. $\mathfrak{M}, \mathbf{x}, V \nVdash Pa$ for some V. Since Pa does not contain any free occurrence of a variable, this amounts to $\mathfrak{M}, \mathbf{x} \nVdash Pa$. Since $\mathfrak{I}(\mathbf{y}, a)$ and $\mathfrak{I}(\mathbf{y}, P)$ are constant for every $\mathbf{y} \in W$, $\mathfrak{M}, \mathfrak{I}(a) \nVdash Pa$. By the definition of $\mathbf{M}^{\mathfrak{M}}$, $I(p_i, g) = 0$. The proofs of the claims for conjunctions and disjunctions are routine and we omit them.

Now assume for some V that $\mathfrak{M}, \mathbf{x}, V \nVdash ST'_a(\alpha \to \beta)$, i.e. $\mathfrak{M}, \mathbf{x}, V \nVdash \Box \forall x (ST'_x(\alpha) \supset ST'_x(\beta))$. This amounts to the claim that there is a $\mathbf{y}$ such that $\mathbf{y} \xrightarrow{\mathbf{x}} \mathbf{z}$ for some $\mathbf{z}$, and $\mathfrak{M}, \mathbf{y}, V \Vdash \exists x(ST'_x(\alpha) \wedge \neg ST'_x(\beta))$. Hence, there is a x-variant V' of V such that $\mathfrak{M}, \mathbf{y}, V' \Vdash ST'_x(\alpha)$ and $\mathfrak{M}, \mathbf{y}, V' \Vdash \neg ST'_x(\beta)$. By the definition of $\mathbf{M}^{\mathfrak{M}}$ and Lemma 7, there is a $w \in K$ such that $I(\alpha, w) = 1$ and $I(\beta, w) = 0$. Hence, $I(\alpha \to \beta, g) = 0$. □

Corollary 9 *If $ST'_a(\alpha)$ is not valid in the class of specific models, then $\alpha \notin \mathsf{B}^+$.*

Now we may state our main result.

Theorem 10 $\alpha \in \mathsf{B}^+$ iff $ST'_a(\alpha)$ is valid in the class of specific propositional models.

Proof. Lemma 5 and Corollary 9. □

6 Conclusion

Theorem 10 shows that there is a relative embedding of B^+ into a version of term-modal logic. To be more specific, there is a translation ST'_a such that $\alpha \in \mathsf{B}^+$ iff $ST'_a(\alpha)$ is valid in a special class of term-modal models. This result may be interpreted philosophically as demonstrating that modal logic can simulate relevant implication.

However, the embedding theorem may be perceived as rather weak. First, we have to confine our attention to a specific class of term-modal models for the embedding to work. There is no immediate solution to this, since the condition that the domain of the model be identical with its set of points is crucial. Second, term-modal logics may be considered by some as a rather specific kind of modal logics. To justify our philosophical claim from the end of the previous paragraph, one may want an embedding into a more "standard" modal logic. Third, the theorem is concerned only with B^+. One may want to obtain similar results for stronger logics (with negation), such as R. Pursuing these lines is left for another occasion.

References

Beall, J., Brady, R., Dunn, J., Hazen, A., Mares, E., Meyer, R., et al. (in press). On the ternary relation and conditionality. *Journal of Philosophical Logic*.

Blackburn, P., Rijke, M. de, & Venema, Y. (2001). *Modal logic*. Cambridge: Cambridge UP.

Dunn, J. M., & Meyer, R. K. (1997). Combinators and structurally free logic. *Logic Journal of the IGPL*, *5*(4), 505–537.

Fitting, M., Thalmann, L., & Voronkov, A. (2001). Term-modal logics. *Studia Logica*, *69*(1), 133–169.

Kurtonina, N. (1995). *Frames and Labels: A modal analysis of categorial inference*. Unpublished doctoral dissertation, Utrecht University.

Kurtonina, N. (1998). Categorial inference and modal logic. *Journal of Logic, Language and Information*, *7*(4), 399–411.
Meyer, R. K. (1975). Relevance is not reducible to modality. In A. R. Anderson & N. D. Belnap (Eds.), *Entailment: The logic of relevance and necessity* (Vol. 1, pp. 462–471). Princeton: Princeton University Press.
Priest, G., & Sylvan, R. (1992). Simplified semantics for basic relevant logics. *The Journal of Philosophical Logic*, *21*(2), 217–232.
Thalmann, L. (2000). *Term-modal logic and quantifier-free dynamic assignment logic*. Unpublished doctoral dissertation, Uppsala University.

Igor Sedlár
Department of Logic and Methodology of Science,
Comenius University
Šafárikovo námestie 6, 818 01 Bratislava, Slovakia
e-mail: `sedlar@fphil.uniba.sk`
URL: `https://sites.google.com/site/sedlarsite/`

Trying to Model Metaphor

Krister Segerberg*

Abstract

Among the many possible functions of metaphor—there are many—one is to convey information. In this short paper I try to outline a theory for what might happen when someone tries to get at the information contained in someone else's metaphor.

Keywords: metaphor, semantics

Fish out of water: a case study

One day I receive a one-line message from my friend John: "Here I am, a fish out of water." How am I to understand this?

John is a professor at a well-known university who has been invited to spend his sabbatical at this reputable think-tank. He should be in his element, one would have thought, but evidently he is not. It turns out that the other fellows are all working in quite different fields. And, like John, they are totally absorbed by their own research interests. Every day there is morning tea, lunch, and afternoon tea at which the fellows meet. And when they meet they talk to one another. But no-one talks to John.

Suppose that in the common room of the think-tank there is an aquarium with some gold-fish. Suppose that one of them, Fred, let us say, suddenly finds himself "out of water". We may assume that Fred is a particularly lively goldfish that enjoys making splashes by jumping up out of the water. But this time he put too much effort into his jump, and something went wrong. So there he is now, thrashing

---

*Most of the work reported here was done at the Goethe University in Frankfurt-am-Main during a visit 2010–11. I am grateful to the Humboldt Foundation for making this visit possible and to André Fuhrmann and his department for making it so pleasant.

on the floor beside the aquarium. John, seeing Fred, identifies with him. "That's me", John says to himself: "a fish out of water". Having found this poignant expression of his predicament, John hastens to communicate it to his friends. (After first putting Fred back into his element, let's hope.)

John and Fred are of course different in many ways:

John is a philosopher.	Fred is a fish.
John cannot swim.	Fred is a superb swimmer.
John is a vegetarian.	Fred is omnivorous.
John loves Wagner.	Fred does not care about music.
John is well-known in his field.	Fred is just an ordinary fish.

Nevertheless, different though they are, their respective situations are similar or at least in some ways parallel. For example:

John needs colleagues to talk to.	Fred needs oxygen.
John has not got enough colleagues to talk to.	Fred has not got enough oxygen.
John is bored.	Fred is dying.

Actually, if the situations of John and Fred are described in more general terms, it appears that they have quite a lot in common:

John is in unfamiliar surroundings.	Fred is in unfamiliar surroundings.
John feels bad.	Fred feels bad.
John has a problem.	Fred has a problem.
John does not know what to do.	Fred does not know what to do.
John wishes he were somewhere else.	Fred wishes he were somewhere else.
John needs help.	Fred needs help.

At this point, let us forget about Fred. The metaphor has had helped us, the hearers, to distill, from the information contained in John's message, a set of propositions that we accept as true in the given situation. In other words, the metaphoric proposition "John is a fish out of water" has given rise to—suggested—a set of informative, non-metaphorical propositions.

John's metaphor is not uncommon. In fact it is so common that one might even try to define it. A dictionary listing locutions of this kind would offer a definition, perhaps something like this:

> **fish out of water** Said of someone who is out of his or her element.

Or more informatively:

> **fish out of water** Occurs in the locution *to be a fish out of water*, referring to someone who is in unfamiliar or uncongenial surroundings.

A dictionary offers only a generic formulation. The statements in the third group of examples above are more specific. But even they do not exhaust all there would be to say. Perhaps no finite list would be enough for that. If so, the metaphor would really be needed; there would be no other way for John to express *exactly* what he did express by describing himself as a fish out of water. Even the formulations in the third group are only approximations. This is where the interest of this paper lies: in "genuine" metaphors—metaphors that cannot be adequately translated into the current language; metaphors that force us to extend our current language.

Two comments

Before moving on to more constructive thoughts, let us make two comments. One is that the metaphors studied in this paper are what, for want of a better term, may be called *simple*. In Lakoff and Johnson's classical treatise *Metaphors we live by* (Lakoff & Johnson, 1980) we meet more complex metaphors as well: metaphors that are systematic in the sense that they can be used to engender an indefinite number of other, "subordinate" or "take-off" metaphors. One of their examples of a metaphor is THEORIES ARE BUILDINGS (to use their own way of denoting metaphors). For this particular metaphor Lakoff and Johnson offer four "take-offs" (Lakoff & Johnson, 1980, p. 53):

- His theories have thousands of little rooms and long winding corridors.

- His theories are Bauhaus in the pseudofunctional simplicity.

- He prefers massive Gothic theories covered with gargoyles.
- Complex theories usually have problems with the plumbing.

The other comment concerns the venerable distinction between 'simile' and 'metaphor'. If John had said that he *was like* a fish out of water, he would have given expression to a simile. But he said, or indicated, that he *was* a fish out of water, thereby expressing a metaphor. A statement of the former kind may well have been literally true, while the latter, although metaphorically true, is literally false. It is presumably differences like these that have led many authors to postulate a big divide between simile and metaphor. From a syntactic point of view, such a division may seem important; from a semantic one, less so; at least for simple metaphors the distinction would seem to vanish. In other words, if one is trying to build formal models of the kind we are interested in here, the problems of modelling "Juliet is like the sun" (a simile) and "Juliet is the sun" (a metaphor) will offer similar challenges. This is of course not to deny that other authors, with other purposes in mind, may have excellent reasons for separating similes and metaphors.

The meaning of a simple metaphor is a convex set

In the analysis of our example in the first section we are dealing with three items: a subject, a copula, and a predicate. Strictly speaking there are at least three alternatives of looking at the matter (the asterisk signals the need for a non-normal reading of the expression in question):

John* is (a fish out of water),

John is* (a fish out of water),

John is (a-fish-out-of-water)*.

The first alternative is the one we actually employed at the beginning of the paper when Fred, the gold-fish, was imagined. But he was, after all, only a thought experiment. Besides, essentialists would hold that it is impossible that John could have been a fish and that therefore this alternative just does not make sense. So let us drop that alternative.

The second alternative suggests "is" does not really stand for ordinary being but rather for something like "is like". But then the metaphor is lost, for to say that John is like a fish in water is to resort to simile, not metaphor.

So we seem reduced to the third alternative: there is a property (fish-out-of-water)*—the metaphorical fish-out-of-water property—and John has that property.

In more technical language, our hypothesis is that, within the abstract model theory proposed below, the "point" of a metaphor can be viewed as a certain "fixed point": a set of propositions that are true in a situation both literally and metaphorically. In our example, that set would consist of the propositions in the third group above (beginning with "John is in unfamiliar circumstances") plus an indefinite number of similar propositions. The question whether a certain metaphor is reducible to (definable in) non-metaphorical language is then the same question as whether that set is finitely axiomatizable. And to repeat our point: a genuine metaphor is not finitely axiomatizable.

In these terms we can formulate a hypothesis. When John claims to be a fish out of water (which obviously he is not), then he is presumably referring to what is common to his own situation and that of a typical, real, living fish that finds itself in a waterless environment. The idea is that he and such a fish share certain properties that can be meaningfully ascribed to each of them. And it is by looking at Fred's situation that we can hope to gain information about John's; for rather than trying to say something directly about his own situation, it is Fred's that John is holding up, presumably as an illustration of his own.

Now, the most striking of the properties that John and Fred share is one for which our ordinary language does not seem to have a term exactly expressing it. This is where the expedient of metaphor comes in, for it does provides us with a term: one that we may name the (a-fish-out-of-water)* property (the asterisk suggesting metaphorhood). With this term our (slightly expanded) language is richer, more expressive than it was without it.

Except what does it mean to be (a-fish-out-of-water)*?

To answer this question we will draw on an interesting suggestion by Peter Gärdenfors (2000, p. 70f.). Here he first quotes a previous author, N. R. Shepard:

An object that is significant for an individual's survival and reproduction is never sui generis; it is always a member of a particular class—what philosophers call a "natural kind." Such a class corresponds to some region in the individual's psychological space, which I call a consequential region. I suggest that the psycho-physical function that maps physical parameter space into a species' psychological space has been shaped over evolutionary history so that consequential regions for that species, although variously shaped, are not consistently elongated or flattened in particular directions.

Gärdenfors's comment on this passage:

Although Shepard does not give any explanation of why evolution should prefer regions that are not oddly shaped, I believe that this can be defended by a principle of *cognitive ecomomy*; handling convex sets puts less strain on learning, on your memory, and on your processing capacities than working with arbitrarily shaped regions.

And so Gärdenfors goes on to make an important suggestion that he himself terms a criterion:

A *natural property* is a convex region of a domain in contextual space.

The terms 'domain' and 'contextual space' are concepts with a certain precise meaning in (Gärdenfors, 2000) that need not concern us here. But we do need to define two topological concepts, those of 'betweenness' and 'convexity'.

Let I be a given set of (possible) individuals with a Stone topology T, and let B be a ternary relation on I. If x, y and z are elements of I we read "y is strictly between x and y" for $B(x, y, z)$. There are the following conditions:

(B1) If $B(x, y, z)$ then $B(z, y, x)$ ("quasi-symmetry"),

(B2) If $B(x, y, z)$ then not-$B(y, x, z)$ & not-$B(z, x, y)$

    ("quasi-antisymmetry"),

(B3) If $B(x,y,z)$ & $B(y,z,w)$ then $B(x,y,w)$ & $B(x,z,w)$
("quasi-transitivity"),

(B4) If $B(x,y,z)$ & $B(y,w,z)$ then $B(x,y,w)$ & $B(x,w,z)$
("tightness"),

(B5) If $B(x,y,z)$ then $x \neq y$ & $y \neq z$ & $x \neq z$ ("strictness").[1]

We say that a subset X of the given space is *convex* if, for all points $x, y \in X$, if w is a point such that $B(x, w, y)$, then $w \in X$. In other words, a set is convex if and only if it includes all points situated between any two points in the set. Note that if X is a subset of I, then there will be a smallest convex set that includes X; this set we call *the convex hull* of X, in symbols $\mathcal{C}X$. We may think of $\mathcal{C}$ as an operator taking subsets of I to subsets of I. The following properties of $\mathcal{C}$ are worth noting: for all $X, Y \subseteq I$,

(C1) $\mathcal{C}(X \cap Y) = \mathcal{C}X \cap \mathcal{C}Y$,

(C2) $X \subseteq \mathcal{C}X$,

(C3) $\mathcal{C}\mathcal{C}X \subseteq \mathcal{C}X$,

(C4) $\mathcal{C}U = U$,

(C5) $\mathcal{C}\emptyset = \emptyset$.

Thus $\mathcal{C}$ is what is known as a *closure operator*.

Let us write F^l for the property of being, literally, a fish out of water, and let us write F^m for the property of being, metaphorically, a fish out of water. In our modelling, we may think of both F^l and F^m as subsets of the space I. (Notice that F^l and F^m are not disjoint; in fact, Fred is a fish out of water both literally and metaphorically.) Let us write

$F^l = \{ i \in I : i \text{ is a fish out of water} \}$ (the literal case)

$F^m = \{ i \in I : i \text{ is (a-fish-out-of-water)}^* \}$ (the metaphorical case)

---

[1] These postulates can be simplified, as they are in (Gärdenfors, 2000, p. 15f.). Note that there is one condition that we do *not* have in full generality:

(B*) If $B(x, y', z)$ & $B(x, y'', z)$ then $B(x, y', y'')$ or $B(x, y'', y')$
("quasi-connexity").

We are now ready to try to suggest an answer the question posed above: what to make of the metaphorical fish-out-of-water property. Here is our technical answer: it is the convex hull of the union of two properties, that of being John and that of being an actual fish out of water. In symbols:

$$F^m = \bigcap \{\, X \subseteq I : X \text{ is convex \& John} \in X \,\&\, F^l \subseteq X \,\}$$

This analysis, sketchy as it is, suggests the same conclusion as that in the preceding section, namely, that the metaphorical fish-out-of-water property coincides with the intersection of all sufficiently general, real properties possessed by an actual fish out of water.

Concluding remarks

The formal representation in the preceding section is just that: formal. In order to connect with the informal analysis given in the first section, one would have to explain the topological connexion. Such an explanation would have to do with the fact that topological terms can be used to express nearness. However, what one would take nearness to be in this particular case cannot be settled by the abstract theory of topology: it is up to speakers and listeners to make their own interpretations. The remarkable thing is that speakers and hearers are able to do this. Apt metaphors are understood by collaborating listeners.

In our example, John and Fred are nearer to one other (their situations are more similar) than some other couples are. For example, John is in some important sense closer to Fred than he is to his colleagues at the think-tank, and similarly Fred is closer to John than he is to his friend Harry who is still happily swimming in his bowl. The topology goes with the perspective.

So when John sends me his laconic S.O.S. message we may think of him as implicitly defining a certain topology. And if I am sufficiently sensitive and sufficiently patient, I should able to work out at least roughly what that topology might look like. But how that may be done is something that is left open in this paper. (A more formal effort to contribute to the analysis of metaphor was made in Segerberg, 2011.)

References

Gärdenfors, P. (2000). *Conceptual spaces: The geometry of thought.* Cambridge, Mass.: The MIT Press.

Lakoff, G., & Johnson, M. (1980). *Metaphors we live by.* Chicago: The University of Chicago Press.

Segerberg, K. (2011). A modal logic of metaphor. *Studia Logica, 99*, 337–347.

Krister Segerberg
Department of Philosophy, Uppsala University, Sweden
e-mail: `krister.segerberg@filosofi.uu.se`

Truth & Knowledge in Logic & Mathematics

Gila Sher

Abstract

In this paper I develop an account of truth and knowledge for logic and mathematic. The underlying methodology is a synthesis of holistic and foundational principles, logical and mathematical truth are based on indirect correspondence, and logical and mathematical knowledge is quasi-apriori (as opposed to being either empirical or purely-apriori).

Keywords: truth, knowledge, logic, mathematics, composite reference, composite correspondence, quasi-apriori, Benacerraf

Logic and mathematics are abstract disciplines par excellence. What is the nature of truth and knowledge in these disciplines? In this paper I investigate the possibility of a new approach to this question. The underlying idea is that knowledge qua knowledge, including logical and mathematical knowledge, has a dual grounding in mind and reality, and the standard of truth applicable to all knowledge is a correspondence standard. This applies to logic and mathematics as much as to other disciplines; i.e., logical and mathematical truth are based on correspondence. But the view that logical and mathematical truth are (i) based on correspondence and (ii) require a grounding in reality demands a change in the common conception of both correspondence and epistemic grounding.

Before turning to this task, however, I have to address the questions (a) Do logic and mathematics, as highly abstract disciplines, require a grounding in reality (or do humans need a logic and a mathematics that are grounded in reality)?, and (b) Is there is anything in reality for logic and mathematics to correspond to? After giving positive answers to these questions, I will turn to the traditional methodology of grounding branches of knowledge in reality ("foundationalism") and the traditional theory of truth associated with correspondence ("copy", "mirror", or "isomorphism" theory), and I will propose an alternative. A foundational methodology for logic and mathematics, I will argue, ought to be holistic rather than foundationalist; but being holistic does not mean being coherentist. On the contrary. Grounding highly abstract disciplines in reality requires a large array of interconnected cognitive resources and a wide network of interconnected routes from mind to reality, and a holistic conception of cognition is better suited to explaining how these requirements are satisfied. I will call the use of holistic methods to pursue the foundational project "foundational holism".

As for correspondence, traditionally correspondence is viewed as based on a single and simple principle, one that assumes the same form in all fields, be they largely observational or highly abstract. However, given the large array of fields and the substantial differences between them (for example, differences in the aspects of reality they study), this view is quite unreasonable. Likewise, the idea that the correspondence between true statements (true theories) and reality is always simple or direct is unreasonable. Instead, I will suggest that mathematical correspondence is "composite" (indirect) and that logical correspondence is closely related to it. I will describe a template of composite correspondence that can be used for mathematics and an associated template for logic, and I will show how these and other elements of the present account equip it to solve some outstanding problems in the philosophy of logic and mathematics. Due to limitations of space I will be able to offer only a general outline and a few examples.

1 Truth in mathematics

Our first question is: If truth in mathematics is based on correspondence, what is mathematical correspondence correspondence with? My proposed answer is that mathematical correspondence is correspondence with the formal layer (aspect, dimension, structure) of reality, the layer of formal features, or formal "behavior", of objects. This answer, however, raises two new questions, the critical question "Does reality have a formal layer or structure?" and the clarificatory question "What do we mean by 'formal'"? Starting with a somewhat vague characterization of the formal as sensitive to the *patterns* of objects having properties and standing in relations but not to the identity or type of objects involved (a more precise characterization will be given shortly), let us turn to the first, critical, question. To avoid a conflict with nominalists right at the beginning, let us think of the ontology of individuals as limited to observable individuals. The question, then, is: Do objects in the world have formal properties? My answer is positive: Individuals (0-level objects) have the formal property of self-identity; properties of individuals (1st-level objects) have cardinality properties, which are formal; relations of individuals (1st-level objects) have formal properties like reflexivity, symmetry, and transitively, and so on. As for a precise characterization of formality, I propose *invariance under isomorphisms* (Lindström, 1966) as capturing the informal idea with which we have started. To see what is meant by "invariance under isomorphisms", let us define for each property a class of argument-structures, i.e., a class of pairs consisting of a universe and an argument of the given property in that universe. An argument-structure of self-identity, for example, is a pair $\langle A, a \rangle$ where A is a non-empty set of individuals and a is a member of A; an argument-structure of a cardinality property is a pair, $\langle A, B \rangle$, where A is as above and B is a subset of A, and an argument-structure of a property like "is symmetrical" is a pair $\langle A, R \rangle$, where R is a binary relation on A. These properties are invariant under isomorphisms in the sense that they are preserved under isomorphisms of their argument-structures: (i) a is self-identical, $\langle A, a \rangle \cong \langle A', a' \rangle \Longrightarrow a'$ is self-identical, (ii) Exactly one individual is B (or all but two individuals are B), $\langle A, B \rangle \cong \langle A', B' \rangle \Longrightarrow$ Exactly one individual is B' (all but two individuals are B'), and (iii) R is symmetric, $\langle A, R \rangle \cong \langle A', R' \rangle \Longrightarrow R'$ is symmetric.

Now, if formal properties are real (in the sense that objects in the world—actual and potential—have, or could have, such properties), then it is reasonable to surmise that such properties are governed by laws. The question arises: Which discipline, if any, studies these laws? Our answer is that mathematics does (or some of its parts do). To see the force of this answer, suppose mathematics does not study the laws governing properties of objects in the world. This would be quite strange. It would be quite strange if arithmetic and set theory studied the laws governing *imaginary* cardinalities but *real* cardinalities were governed by altogether different laws. This does not mean that the only thing that mathematics does is study the formal laws governing the behavior of objects and structures of objects in the world, but that one important thing it does is study these laws.

Here, however, we come upon a puzzle: It appears that the formal features of objects in the world are for the most part of a relatively high level (2nd-level and above), but mathematical theories are for the most part 1st-order theories. Why do arithmetic and set theory study cardinalities as *individuals* (the numbers zero, one, two, ...) if they are 2nd-level properties (ZERO, ONE, TWO, ...)?

My answer is that there could be many reasons why people prefer to think of cardinalities as individuals and mathematicians prefer to study them by 1st-order theories. For example, it is quite possible that we, humans, work better with individuals and their properties than with properties and their properties. It is easier for us to figure out, and present in a systematic manner, the laws governing cardinalities when we think of cardinalities as individuals than as 2nd-level properties. The main issue is: Is it possible to account for 2nd-level phenomena *accurately* and *systematically* by 1st-order theories?

The answer to this question is quite clearly positive. Mathematics may study cardinalities indirectly yet accurately and systematically. Before we further elaborate on this answer, let us reflect, more generally, on some of the factors involved in humans' ability (or inability) to reach the world cognitively and develop a standard of truth for theories (statements, thoughts, beliefs) about it. Four such factors are:

(a) The complexity of the world.

(b) Our desire to know and understand it in all its complexity.

(c) The mind's cognitive limitations.

(d) Its intricate cognitive capacities.

These factors introduce some tensions into the two projects of theorizing (thinking) about the world and constructing a truth standard for our theories (thoughts) about it. But they also point to a solution. In particular, they suggest that we approach these projects with the following principles as guidelines:

(a) Seek a fruitful balance between unity and diversity. (Dyson, 1988, p. 47)

(b) Access reality holistically.

The first principle suggests that instead of either radical monism or radical pluralism with respect to truth, we adopt a family of diverse yet unified standards of truth. E.g., instead of either a single and simple correspondence standard of truth or radically divergent standards of truth for different fields (correspondence for physics, coherence for mathematics), we allow a *family* of correspondence standards that can take into account the special needs of different fields. The second principle says that in accessing reality we are free to use all the resources available to us (including those involving non-vicious circularity), and that using these resources we are free to forge multiple routes to reality, including multiple correspondence routes. Correspondence, on this conception, is not a "mirror" or a "copy" or even an "isomorphism" relation between language and reality, but rather a family of interrelated connections between the two. These connections enable us to say "of what is that it is and of what is not that it is not" in an accurate, if at times circuitous (indirect, composite, multi-staged), ways. As a standard, correspondence sets substantial yet flexible conditions on our discourse, requiring an appropriate truth-conferring connection between our discourse and reality, given our resources, reality, and what aspect(s) of reality we are speaking about. (The exact content of these conditions may change in the course of history, reflecting changes in our resources, our understanding of reality, and what aspects of reality we are interested in.)

Turning back to mathematics, our question was whether it is possible to accurately account for the formal aspect of reality, which is largely higher-level, by 1st-order mathematical theories. It should now be clear why a positive answer to this question is, in principle, justified. One way in which we can account for 2nd-level phenomena through 1st- and 0-level thoughts is by introducing an intermediate layer of posits into our picture of reality. For example, we can create a posited layer of mathematical individuals and their properties (levels 0 and 1) that systematically represents 2nd-level cardinality properties and their (3rd-level) properties. Mathematical correspondence will then be circuitous yet accurate. Starting with *reference*, we can display the difference between simple and "composite" reference as follows:

Simple Reference	Composite Reference
Lang.: Ind. Consts[0] Predicates[1]	**Lang.:** Ind. Consts[0] Predicates[1]
	↓ ↓
	Posit: Individuals[0] Properties[1]
↓ ↓	↓ ↓
World: Individuals[0] Properties[1]	**World:** Properties[2] Properties[3]

Turning to correspondence, consider the 1st-order arithmetic truths "$2 + 7 = 9$" and "$(\forall m)(\forall n)(m + n = n + m)$". We can describe their composite correspondence-conditions as follows:

Composite Mathematical Correspondence

1st-Order Language: "$2 + 7 = 9$" is true

iff

Posits: $+(2, 7) = 9$

iff

Reality: DISJOINT-UNION(TWO, SEVEN) = NINE

[*iff*

$(\forall P1)(\forall P2)((\text{TWO}(P1) \ \& \ \text{SEVEN}(P2) \ \& \ P1 \cap P2 = \emptyset] \supset \text{NINE}(P1 \cup P2))$]

1st-Order Language: "$(\forall m)(\forall n)(m + n = n + m)$" is true

iff

Posits: The operation of addition is symmetric

iff

Reality: DISJOINT-UNION IS SYMMETRIC.

Three advantages of the composite correspondence principle of mathematical truth together with the other alethic and epistemic principles delineated above are:

(a) They enable us to provide a substantive account of mathematical truth.

(b) They enable us to integrate this account in a unified correspondence account of truth.

(c) They lead to solutions to outstanding problems in the philosophy of mathematics (the large ontology problem, the identity problem, and the applications problem), as well as to progress toward solutions to other problems (the cognitive access problem and the "mathematics as algebra" problem). This will be discussed in Section 4 below.

2 Truth in logic

If truth is correspondence throughout, then logical truth, to the extent that it is a genuine type of truth, is also based on correspondence. Similarly, the other semantic properties (relations) associated with logical truth—logical consequence, logical consistency, etc.—are based on principles related to correspondence. This means that logic has as much to do with the world as it does with the mind (language, concepts, etc.), and the first question we are facing is, therefore: What does logic have to do with the world?

My answer is:

(a) Logic has to "work" in the world. This point was already noted by Russell. Speaking about logical laws (the "*law of identity*", the "*law of contradiction*", the "*law of excluded middle*"), Russell says that "what is important [with respect to these laws] is not the fact that we think in accordance with [them], but the fact that **things behave in accordance with them**". It is this that is responsible for "the fact that when we think in accordance with them we think *truly*". (Russell, 1959, p. 34, my bolding)

(b) There have been cases of *factual* error in logical theories, and some say there still are. The most dramatic example of a discovery of a factual error in a logical theory is the discovery of an error in Frege's logic by Russell. Frege's logic, Russell discovered, affirms the existence of a set that does not, and cannot, exist. Moreover, advocates of so-called nonclassical logics (fuzzy logic, quantum logic, etc.) claim that classical logic fails to work in the world and this is naturally understood as due to errors in understanding the formal structure of reality.

(c) Logic is both constrained and enabled by reality through its inherent connection with truth. (This is a theoretical rendition and expansion of a point noted briefly in the citation from

Russell above.) Consider the case of logical consequence, $\Sigma \models \sigma$, where the truth-conditions of both σ and the sentences in Σ are based on correspondence principles. Say, σ is true iff the situation $\mathfrak{E}_2$ is the case, and the sentences in Σ are true iff $\mathfrak{E}_1$ is the case. For σ to be a logical consequence of Σ, the truth of the sentences in Σ (assuming they are all true) has to be transmitted to (or preserved by) σ with an especially strong modal force. But because the truth of σ and the sentences in Σ is a matter of whether $\mathfrak{E}_2$ and $\mathfrak{E}_1$ are the case, $\mathfrak{E}_1$ must be connected to $\mathfrak{E}_2$ with an especially strong modal force as well. The relation between logical consequence and reality is, thus, as in the following diagram:

$$\begin{array}{ll} \textit{Logic:} & \Sigma \models \sigma \\ & \updownarrow \\ \textit{Truth:} & T(\Sigma) \Rightarrow T(\sigma) \\ & \updownarrow \\ \textit{World:} & \mathfrak{E}_1 \rightarrow \mathfrak{E}_2, \end{array}$$

where $\Rightarrow$ and $\rightarrow$ have an especially strong modal force.

This relation means that the world both constrains and gives rise to logical consequences, or that the world can both falsify and justify logical-consequence claims. Starting with falsification, the point is that we cannot choose a logical theory without paying attention to the world. In particular, if our chosen logical theory says that $\Sigma \models \sigma$ but in the world one of the following three situations is the case:

(a) $\mathfrak{E}_1$ but *not* $\mathfrak{E}_2$,

(b) $\mathfrak{E}_1$ and $\mathfrak{E}_2$, but no law connects $\mathfrak{E}_1$ to $\mathfrak{E}_2$,

(c) $\mathfrak{E}_1$, $\mathfrak{E}_2$, and some law connects $\mathfrak{E}_1$ to $\mathfrak{E}_2$, but this law is not sufficiently strong for logical consequence,

then our chosen theory is wrong: σ is *not* a logical consequence of Σ.

One way in which the world can give rise to logical consequences or support (provide positive evidence) for logical-consequence claims is:

(d) $\mathfrak{E}_1$, $\mathfrak{E}_2$, and $\mathfrak{E}_1$ is connected to $\mathfrak{E}_2$ by a law whose modal force is sufficient for logical consequence.

The questions arise: What modal force is sufficient for logical consequence? Are there laws possessing such modal force? My proposal is that the modal force sufficient for logic is that of formal laws (laws that hold in all formally possible states of affairs), and that there are such laws, namely those studied (directly or indirectly) by mathematical theories like arithmetic and set theory. We have already seen that objects in the world have formal properties and that it is reasonable to surmise that these properties are governed by laws. We have also seen that arithmetic and set theory study these laws. An example of such a law and the logical claim it grounds is:

World: UNIVERSALITY OF $\mathcal{A}$ → UNIVERSALITY OF $\mathcal{A} \cup \mathcal{B}$
$$\Downarrow$$
Truth: $\quad T[(\forall x)Ax] \Rightarrow T[(\forall x)(Ax \vee Bx)]$
$$\Downarrow$$
Logic: $\quad (\forall x)Ax \models (\forall x)(Ax \vee Bx),$

where $\mathcal{A}$ and $\mathcal{B}$ are the properties denoted by "A" and "B", respectively. The logical form of sentences standing in the relation of logical consequence is determines by their (linguistic) formal parameters ("$\forall$" and "$\vee$" in the example above), and these parameters are formal in virtue of representing the (objectual) formal parameters of the situations corresponding to these sentences (UNIVERSALITY and $\cup$).

Our account unites logic with mathematics in a way that is in some respects similar to *logicism*. Like logicism, it unifies the two disciplines, thereby replacing two foundational tasks—the task of grounding logic and the task of grounding mathematics—by one. However, unlike logicism, it provides a grounding not just for one of these disciplines (logicism grounds mathematics in logic leaving logic itself ungrounded), but for both: both logic and mathematics are grounded in the formal. On this conception, the relationship between logic and mathematics is a back and forth relationship: we use our mathematical knowledge (knowledge of the formal) to construct our logical system, and we use our logical system as a framework for constructing mathematical theories. (Think of the mutual relationships between 1st-order logic and set theory, each having played an important role in the development of the other.)

Under this conception, logical constants are referring constants and logical truths are based on correspondence with formal laws governing reality. Logical reference and correspondence have two dimensions: a simple, 1-step dimension, in which they are directly connected to certain features of reality, and a composite, 3-step dimension, in which they are connected to the same features of reality indirectly through mathematical reference and correspondence. The former dimension is *ontological*, the latter—*epistemic*. Logical laws are grounded in largely *higher-level formal laws*, but these laws are known to us through *lower-level mathematical theories* and *lower-level posits*. An example of these two dimensions is:

Simple Logical Reference

Logical language: "$\exists$" (2nd-order predicate serving as 1st-order quantifier)

$$\Downarrow$$

World: (2nd-level property of) NON-EMPTINESS

Simple Logical Correspondence

Logical language: "$\sim(\exists x)(Px \;\&\; \sim Px)$"

$$\Updownarrow$$

World: EMPTY[$\mathcal{P} \cap$ COMPLEMENT($\mathcal{P}$) IN $\mathcal{A}$], where $\mathcal{A}$ is a given universe

But to *know* this and more complex formal laws governing the world, logic turns to mathematics, which establishes them, often indirectly, as 1st-level laws. Therefore, in the order of knowledge, logical reference and correspondence may very well be indirect:

Composite (3-layered) Logical Reference

Logical language: "$\exists$"

$$\Downarrow$$

Mathematical language: "> 0" / "$\neq \emptyset$"

$$\Downarrow$$

Posit: being larger than (the individual) 0 / being different from (the individual) $\emptyset$

$$\Downarrow$$

World: NON-EMPTINESS (of properties of individuals)

Composite (3-layered) Correspondence

Logical language: "$\sim(\exists x)(Px\ \&\ \sim Px)$"
$$\Updownarrow$$
Mathematical language: "$p \cap \text{complement}(p) = \emptyset$"
$$\Updownarrow$$
Posit: Empty[intersection(p, complement(p) in A)]
$$\Updownarrow$$
World: EMPTY[$\mathcal{P} \cap \text{COMPLEMENT}(\mathcal{P})$ in $\mathcal{A}$]

3 Knowledge in logic & mathematics

Logical and mathematical knowledge, on the present proposal, has several distinctive features:

(a) It is grounded in reality, yet it is holistic.

(b) It is grounded in the same reality that scientific knowledge is grounded in, yet in different features of this reality. It follows that there are no two separate realities, a Platonic reality that grounds the abstract sciences (including logic and mathematics) and a mundane reality that grounds the empirical sciences, but there is a single reality with diverse features that grounds both.

(c) The grounding of logical and mathematical knowledge in reality can be either direct or indirect (with the possibility of various combinations of the two).

(d) Logical and mathematical knowledge is grounded not just in reality but also in the mind. Its grounding in the mind is both passive and active. On the one hand it is both constrained and enabled by our biological, psychological, cultural and other resources; on the other hand we are always free to devise new ways of expanding our knowledge and improving its quality: new tests, new forms of evidence, new concepts, etc.

(e) Logical and mathematical knowledge is neither purely apriori nor empirical in the sense of relying primarily on sensory perception. Instead, it is *quasi apriori*. We can represent the contrast between our conception and the traditional conception by a line with five regions:

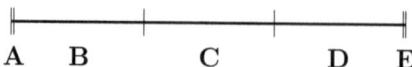

A is the (almost infinitesimally) narrow space of absolutely sensory cognition, i.e., cognition for which not even the slightest influence of reason is permitted; E is the (almost infinitesimally) narrow space of absolutely reason-based cognition, i.e., cognition for which not even the slightest influence of sensory experience is permitted; and B–D are the large intervals of cognitions that can in principle be based both on sensory perception and on reason in various combinations and in a gradually increasing ratio of the latter to the former. In traditional epistemology the *apriori-aposteriori* division has on one side the (almost infinitesimally) narrow region E and on the other side essentially the whole line of knowledge, from A to (and including) D. This is a very uneven division and, moreover, it is a division with sharp boundaries. Logical and mathematical knowledge fall on the (almost infinitesimally) narrow side of this division—it is *purely apriori*, i.e., strictly limited to E. As such its resources are very limited in the sense that no combination of experience and reason is available to it. In contrast, aposteriori, or *empirical* knowledge has a wide array of resources, being entitled (in principle) to *all* combinations of sense-based and reason-based cognitive elements. Given this uneven distribution of cognitive resources between logical and mathematical knowledge on the one side and scientific knowledge on the other, it is not surprising that the task of explaining logical and mathematical knowledge within traditional epistemology is especially difficult and has led philosophers to resort to extreme measures. Two such measures are (i) postulating a separate abstract (Platonic) reality and (ii) viewing logical/mathematical knowledge as (purely) conventional (thereby giving up on genuine truth in these fields).

On the present (foundational holistic) account, regions A and E are eliminated as epistemically significant regions. Human knowledge consists of three continuous (i.e., not sharply divided) regions: the region of largely observational knowledge—B; the region of more theoretical scientific knowledge—C; and the region of highly abstract knowledge—D. Logical and mathematical knowledge resides in D. Due to the special nature of its subject-matter, the *formal*, such knowledge is *quasi-apriori*. I.e., its use of cognitive resources is characterized by a relatively high ratio of reason-based to sensory resources. But we do not rule out cases in which empirical considerations make a significant contribution to changes in mathematics and logic. Among other things, we allows that occasionally empirical discoveries point beyond themselves to phenomena that are highly abstract in nature.

4 Solution to outstanding problems in the philosophy of logic and mathematics

Our account offers a solution to a number of outstanding problems in the philosophy of logic and mathematics, and makes significant steps toward a solution to others.

4.1 Problems in the philosophy of mathematics

4.1.1 The large ontology problem

The problem is to explain the large ontology of mathematics. This is thought to be especially difficult if mathematics is true about the world since the number of individuals in the world appears to be much smaller than that required by mathematics.

On our account, the large ontology of mathematical individuals is an ontology of posits and as such need not be limited to the ontology of real individuals (whatever this is). As for the question, "Why does a theory of the formal need a very large posited ontology?", our answer is: Laws in general are counterfactual in scope, and as such hold for, and may be best formulated in terms of, a counterfactual ontology that, being counterfactual, might be larger than the "actual" ontology. Due to the especially high degree of invariance of formal properties, formal laws have an especially large counterfactual scope; hence they require an especially large posited ontology. For example: to express the laws of finite cardinalities in complete generality, we need an ontology on the order of the denumerable ontology of the natural numbers, and to express the formal laws governing a denumerable ontology (laws like the law of power-set cardinality) we need an ontology on the order of the indenumerable ontology of ZFC.

4.1.2 The identity problem (Benacerraf, 1965)

The problem, as formulated by Benacerraf through an example, is: Zermelo's 2 is $\{\{\emptyset\}\}$, von Neumann's 2 is $\{\emptyset, \{\emptyset\}\}$; which is the real 2?

Our solution to this problem is straightforward: Since 2 is a posit representing the 2nd-level property TWO, it does not matter whether we construe it as $\{\{\emptyset\}\}$, using Zermelo's method, or as $\{\emptyset, \{\emptyset\}\}$, using von Neumann's method (so long as we are consistent). The two are different but equally good representational methods.

4.1.3 The application problem

The problem is how results concerning mathematical objects, which are highly abstract, can apply to physical objects, which are not abstract.

This question is especially difficult for radical Platonists, who believe in two disconnected realities, an abstract reality and a physical reality, and for radical conventionalists, who believe that mathematics is purely conventional and as such disconnected from reality. But on the present account this problem does not arise: physical objects have formal properties, and therefore laws governing these properties apply to them (directly or indirectly) in an unproblematic way.

4.1.4 The cognitive-access problem (Benacerraf, 1973)

Benacerraf's cognitive-access problem is the problem of how, given that our only access to reality is *causal*, we have access to mathematical objects, which do not stand in causal relations to us (or to anything else, for that matter).

Our response has two parts. First, we question the assumption that humans' *only* access to reality is causal; second, we point to the existence of a large network of routes from mind to reality (combining both causal and non-causal elements) that were not considered by Benacerraf. Both points are based on our foundational holistic methodology. The access question has so far been asked from two perspectives, a foundationalist perspective and an anti-foundational perspective, but not from the perspective of foundational holism. Foundational holism, with its rich and multi-layered conception of cognitive access to reality and its view of mathematics as *quasi-apriori*, questions the exclusiveness of causal access to reality, and suggests other ways of accessing it.

4.1.5 The "mathematics as algebra" problem

If mathematics is a theory of the formal, what is the status of algebraic theories, theories that seem to study structures for their own sake?

The present approach points to two ways in which this problem can be dissolved. First, mathematics is a broad and multi-faceted discipline with a variety of interests. One of its central interests is a theory (or theories) of the formal (or of various aspects thereof), but it has other interests as well. Second, algebraic theories offer potential *models* of phenomena in the world (as it is or as it could have been), i.e., models of some states of affairs, but not all. The difference between, say, arithmetic and group theory is, then, this, that while in all formally-possible states of affairs properties of individuals have cardinalities, not all formally-possible structures are group structures.

4.2 Problems in the philosophy of logic

The situation in the philosophy of logic is quite different from that in the philosophy of mathematics. While philosophers of mathematics have, for a long time, been aware of the need to establish the veridicality of mathematics, explain how mathematics works in the world, and clarify the nature of mathematics in a systematic and theoretical manner, philosophers of logic have often neglected these tasks with respect to logic. One exception is Russell who, as we have noted above, was fully aware of the importance of the questions of veridicality and "how it works in the world" for understanding logic. In addition, Russell was fully aware of the need for, and the difficulty of, explaining the nature of logic:

> The fundamental characteristic of logic, obviously, is that which is indicated when we say that logical propositions are true in virtue of their form. ... I confess, however, that I am unable to give any clear account of what is meant by saying that a proposition is "true in virtue of its form". (Russell, 1938, p. xii)

The present paper has offered a solution to these questions, if only in an outline form. (For discussion of related issues see Sher, 1991, 2008, in press.)

References

Benacerraf, P. (1965). What numbers could not be. *Philosophical Review*, *74*, 47–73.

Benacerraf, P. (1973). Mathematical truth. *Journal of Philosophy*, *70*, 661–680.

Dyson, F. (1988). *Infinite in all directions*. New York: Harper & Row.

Lindström, P. (1966). First order predicate logic with generalized quantifiers. *Theoria*, *32*, 186–95.

Russell, B. (1938). Introduction to the second edition. In *The principles of mathematics* (2nd ed., pp. v–xiv). New York: Norton.

Russell, B. (1959). *The problems of philosophy*. New York: Oxford.

Sher, G. (1991). *The bounds of logic: A generalized viewpoint*. Cambridge: MIT.

Sher, G. (2008). Tarski's thesis. In D. Patterson (Ed.), *New essays on Tarski and philosophy* (pp. 300–339). Oxford.

Sher, G. (in press). Forms of correspondence: The intricate route from thought to reality. In C. D. Wright & N. J. Pedersen (Eds.), *Alethic pluralism: Current debates*. Oxford. (Forthcoming)

Gila Sher
Department of Philosophy, University of California, San Diego
La Jolla, 92093-0119
e-mail: gsher@ucsd.edu

www.ingramcontent.com/pod-product-compliance
Lightning Source LLC
Chambersburg PA
CBHW051038160426
43193CB00010B/989